War of Time

Jan Hanska

War of Time

Managing Time and Temporality in Operational Art

palgrave
macmillan

Jan Hanska
Faculty of Social Sciences
University of Lapland
Rovaniemi, Finland

ISBN 978-3-030-45516-3 ISBN 978-3-030-45517-0 (eBook)
https://doi.org/10.1007/978-3-030-45517-0

This Palgrave Macmillan imprint is published by the registered company Springer Nature Switzerland AG.
The registered company address is: Gewerbestrasse 11, 6330 Cham, Switzerland

To my hero, my idol, my confidant, my friend—my father
K.-J. Hanska

Contents

About the Author

Jan Hanska is a general staff officer serving in the Finnish Defence Forces and an adjunct professor of International Relations and War Studies at the University of Lapland. He has received the degree of Doctor of Social Sciences from the University of Tampere and the degree of Doctor of Military Studies from the National Defence University.

CHAPTER 1

Theoretical Grounding for Temporality in Warfare

1.1 Introduction

> The further back one goes, the less useful military history becomes, growing poorer and barer at the same time. The history of antiquity is without doubt the most useless and barest of all. This uselessness is of course not absolute; it refers only to matters that depend on precise knowledge of the actual circumstances, or on the details in which warfare has changed. (Clausewitz 1989, p. 173)

One should not unquestioningly adopt this maxim by Carl von Clausewitz. Military history of the antiquity is useless only if one searches for detailed information on execution of battles. It is beneficial for students of the art of war who seek to penetrate the veil of time and see what could be learned from antiquity and how these lessons could be adopted and adapted to suit warfare of our times. As Dupuy (1987, p. xxii) argued, war is an observational science like astronomy. For the soldier the best laboratory is military history. Russian military tradition has relied heavily on historical operational analyses. Analytical study of the past was considered essential for predicting future developments in warfare (Adamsky 2010, p. 33). As Svechin (1992, p. 279) put it, "Work on military history and the art of war can improve our capabilities of drawing up good plans." For Napoleon (1987, p. 81–82) tactics could be learned from treatises, but knowledge of strategy is

J. Hanska, *War of Time*,
https://doi.org/10.1007/978-3-030-45517-0_1

> acquired by experience, and by studying the campaigns of all the great captains. Gustavus Adolphus, Turenne, and Frederick, as well as Alexander, Hannibal, and Caesar, have all acted upon the same principles. These have been: to keep their forces united; to leave no part unguarded; to seize with rapidity on important points.

If one cannot learn command of armies in *praxis*, the other option is to peruse how the "great captains" led their troops. The aspiring commander needs to model himself upon these examples and hopefully "learn to reject all maxims foreign to the principles of these great commanders." (Napoleon 1987, p.82). Napoleon didn't mean that war should be carried out by copying past history. His argument was that the basic principles of warfare remain intact.

This book focuses on the importance of meaning and management of time in the art of war and especially operational art. Theorists and pragmatists from different cultures and eras have been brought together to compare the roles of time and timing concerning the conduct of war in their writings. Views on warfare have greatly varied among different cultures. The art of war in Russian thought has traditionally been more of an 'arithmetic' type, based on calculations and strict rules. The German thinkers have tended to rely on overwhelming force to produce Wagnerian symphonies. The Greco-Roman culture used heavy tools to sculpt their military masterpieces and if the Oriental art of war reads like simplistic but deep poems of *haiku,* the French and the British have written their romances with the blood of their enemies and their own soldiers alike. There are numerous approaches to the art of war and this book will examine time as a factor in operational art. To quote Clausewitz again:

> When the strength and capability of armed forces are being calculated, time is apt to be treated as a factor in total strength on the analogy of dynamics. It is assumed in consequence that half the effort or half the total forces could achieve as much in two years as the whole could do in one. This assumption, which rests, sometimes explicitly, sometimes implicitly, at the basis of military planning, is entirely false. Like everything else in life, a military operation takes time. No one, obviously, can march from Vilna to Moscow in a week; but here there is no trace of that reciprocal relationship between time and energy that we could find in dynamics. Both belligerents need time; the question is only which of the two can expect to derive special advantages from it in the light of his own situation. (Clausewitz 1989, p. 597)

Time has a profound meaning in the dynamics of warfare but the relationship is more convoluted than would initially seem. Twice the forces, twice the speed is no formula for success. While since the days of Clausewitz it has become a matter of hours to transport troops from Berlin to Moscow, it remains true that every military operation has duration. As Vego (2009, p.III–58) writes, "Any movement in space requires time. Obviously, the longer the distance, the longer the time required to overcome the factor of space. Also, the greater the speed and mobility, the shorter the transit time." This type of thinking about time and space was characteristic to the industrialized age. But especially in our information age, it is insufficient for any commander to satisfy himself with tackling only the problems related to the speed of movement of troops, since every activity requires time. Planning, decision-making, issuing orders, and executing given orders consume the time at the disposal of the commanders, their staffs, and their subordinates.

There are many maxims concerning proper timing in the art of war and operational art. Some remain constant as times change and some become obsolete as the ways and means of waging war develop. Leo VI (2010, p. 635) advised his general "to embark upon your expeditions at that time of the year when the harvest is ready." This maxim is a product of its time and related to a certain type of army and its means of provisioning. It refers to an age when armies foraged and provisioned food from the civilian population and thus full granaries were essential. Such guidance suits only a relatively small army, since no region is agriculturally rich enough to feed a million-man mass army. As times change warfare follows. Therefore, old maxims have to be evaluated anew before they are accepted to shape doctrines. The principles of war don't remain unaltered from one age of warfare to another. If two armies of equal strength, training, equipment, and mobility face each other, they likely need approximately the same amount of time to carry out their operations. Thus it remains a race in which one can gain the "special advantages" from his disposition and maximize the gains by utilizing time as effectively as possible. Every attempt must be made to win more time for oneself and deprive the enemy of his.

One of the reasons for writing this book comes from Liddell Hart's idea of how one might understand war today and perhaps even envision its future outlook. The war of tomorrow is connected to the present way of war and this, in turn, is determined by the past. Evolution of warfare is a continuum.

> The future is moulded by the past. The best promise for the future lies in understanding, and applying, the lessons of the past. For that reason, in discussing the problems created by the current war, more light may come from tracing the whole course of the revolution in warfare than by dealing merely with the appearances of the moment. If we realize how the conditions of this war have come about, there may be some prospect of averting a more deadly recurrence. (Liddell Hart 1946, p. 76)

Many proponents of revolution in military affairs (RMA), information war, network-centric war, cyberwar, and numerous others have fallen into this pit of isolating how war could technically be fought at a given time from the way it has been fought in the past. In the U.S. the RMA was seen as a revolution that consisted of technological drivers that created a shift from brute force to brain force (Strachan 2013, p. 47). The same hype has returned time after time. The outlook of warfare is determined by the level of technological and intellectual sophistication of the society that wages it, but the opponent may be on another level and thus abides to his own rules and worldview. An information society cannot effectively wage an information war with the entire spectrum of its high-tech arsenal against an agrarian society (Kagan 2006, p. 319–322). The most sophisticated cyber weapon is useless if the enemy lacks computers. Furthermore, there have been numerous occasions when one or all proponents have reverted during the course of a given war to ways and means of warfare of the past that may have seemed barbaric at the time.

Ardant du Picq (1987, p. 130) argued that "The study of the past alone can give us a true perception of practical methods, and enable us to see how the soldier will inevitably fight tomorrow." Even during the course of a single war warfare may change drastically. There is a continuous evolution of weapons and methods alike, but in a prolonged war high-technology weapons and other resources may be depleted and this affects the outlook of the war. Reading the famous military thinkers, one notices that many of them considered themselves as witnesses of revolutionary turning points. An example is Giulio Douhet's (1999, p. 383) claim that "in the period of history through which we are passing, war is undergoing a profound and radical change in character and forms, as I shall show; so that the war of the future will be very different from all wars of the past." For the authors their eras were separated from the past.

The fascinating thing about all turning points of history is that people at the time rarely recognize the importance of the moment. Warfare has its own turning points that shape the outlook of future wars. For Douhet (1999, p. 279), "the form of any war—and it is the form which is of primary interest to men of war—depends upon the technical means of war available." We might or might not today live in a time that could in the future be characterized as a turning point—maybe the dawn of robotic and autonomous systems' warfare. Or something completely different. Progress keeps speeding up and we need to keep up with the pace of "mechanical progress, whereby the latest product of to-day is obsolete to-morrow. (…) if we are to await mechanical finality we shall wait for all eternity" (Liddell Hart 1927, p. 13). Development will never stop as long as civilization prospers but its direction remains unfathomable.

> The battle of machines, is this the ultimate goal in warfare? I do not think so, for a machine is but a means of waging war, a tool whereby men seek to impose their will upon each other. (…) Yet, whatever it be, it is the will and understanding of man which the machine forces man to accept. (Fuller 1923, p. 168)

Yesterday, today, and tomorrow, the deciding factors in warfare will be the intelligence, determination, will, and imagination of the human mind. Machines are tools and humans allocate their tasks. As we move from one age into another the outlook of warfare will inevitably change, but the essence of war remains immutable. The acme of skill in the art of war is to "comprehend in good time the momentarily changing situation and after that to do the simplest and most natural things with steadfastness and prudence" (Moltke 1993, p. 175). Time and timing are important in operational art and some ideas remain valid from one age into another, while others need to be discarded or modified to better suit the context. We don't globally live in the same age. While some societies experience full bloom of information age, their neighbors may remain industrial or rooted to the agrarian age. Furthermore, different strata in these societies may exhibit characteristics of different ages. One's own armed forces and those of the enemy may abide to different conceptions of temporality.

Throughout this book 'war' is used to refer both to war as a phenomenon and a period of time that cannot be described as 'peace' even if actual fighting ceases for prolonged periods. War refers both to a state of conflict that surpasses the boundaries of traditional political intercourse between states and a clash of arms between states of other conflicting parties. War is a legal state between belligerents in which international laws of war apply to recognized institutions in international relations. Combat is not a necessity for the state of war to prevail and war doesn't have to be officially declared. 'Warfare' means the actual conduct of war and how the belligerents carry out their warlike activities principally in the military dimension (Gray 2007, p. 32). Here it refers to the praxis of war during active operations. 'Art of war' is used as overarching concept embracing the mental and intellectual aspects of both thinking about war and the practice of warfare on all levels. 'Operational art,' however, is a more tightly defined 'subspecies' of the art of war. The realm of operational art occupies the intermediate area between strategy and tactics, partially overlapping the two and connecting tactical level battles into operations to influence war on the strategic level. Traditionally the term 'campaigning' was used to cover much of what we now consider to be operations and today some favor discussing 'theater-level' instead of operational level, but this is semantics (Smith 2008, p. 13–17). Again, 'operations' deals with the practice of campaigns while operational art focuses more on intellectual and mental aspects of conceiving, planning, and thinking about warfare on the operational level. Hew Strachan (2013, p. 210) defined the operational level as "the level of command situated between the tactical and the strategic, between the company or battalion commander in the field and the president in the White House." On the operational level strategic tasks are broken down into missions that tactical units can accomplish (Kagan 2006, p. 60).

For the purpose of comprehensiveness, it is useful to occasionally discard the thought of dividing the multifaceted and multi-layered concept of warfare into different levels. Time is involved in every military activity on all levels. The brigade commander may strike with a battalion or a nation-state with its air force. Fundamentally the manipulation of time to one's own benefit is similar. On all levels time should be managed, controlled, and exploited while the amount on time and methods may vary. Clausewitz (1989, p. 132) argued that "tactics and strategy are two activities that permeate one another in time and space but are nevertheless essentially different. Their inherent laws and mutual relationship cannot be understood without a total comprehension of both." A battle with all its tactical decisions is simultaneously a part of the big picture, the operational and strategic realms of war. As Jomini (2007, p. 131) wrote, battles are not isolated incidents without a part in the outcome of war. "Battles are the actual conflicts of armies contending about great questions of national policy and of strategy." Battles are practical tools of politicians and strategists during war. Through them, the policy of nation is enacted if other means prove insufficient.

This book will not discuss in detail individual battles and campaigns, but the mindset of theorists and practitioners toward their art. On the upcoming pages there will be references to specific battles but no in-depth depictions. Generalizations may be drawn, but as Ferdinand Foch (1920, p. 11) wrote, "in war there are none but particular cases; everything has there an individual nature; nothing ever repeats itself." The discussion focuses on how the military thinkers depicted their battles with subsequent lessons learned and not what historically happened.

The literature used is three-fold. First group of authors consists of renowned theorists, like B.H. Liddell Hart, who held only the rank of captain, but wrote extensively on the art of war. Major-general J.F.C. Fuller is another example of a theorist without relevant hands-on command experience. Second group includes practitioners of war, men often less theoretical, but with extensive knowledge and experience on command and decision-making in warfare. The memoirs of these generals and field marshals help interpret their ways of thought concerning war. A third group consisting of military thinkers and doers bridges the two above-mentioned. Among this group belong those rare and fortunate ones like Heinz Guderian, who both played a role in formulating operational art and had the opportunity to practice it as commanders.

I have chosen *artistes*, theorists, and practitioners of war based on the influence they have had either in their own time or throughout centuries. Commanders are not always highly esteemed for their literary efforts. As Wellington wrote in a letter, "I am really too hard worked to become an Author and review these lying works called Histories" (cited in Bassford et al. 2010, p. 47). Thus, valuable sources like Alexander, Cyrus, or Hannibal are unavailable to the researcher. When using old theorists to answer contemporary dilemmas, one needs to view them as men writing for their times. The further back in time one goes and the wider the gap between the cultures of the author and his interpreter, the more pressing the need to differentiate between universal and context-specific ideas.

Actual doctrines of the armies are excluded. Liddell Hart (1932, p. 238) argued that official manuals due to their nature "are merely registers of prevailing practice, not the log-books of a scientific study of war." Furthermore, doctrine doesn't equal operational art. Earlier doctrines were associated tightly with the tactical level. They were close to dogma, teaching the soldiers how to fight. In the twenty-first century doctrine approaches the strategic level of war and attempts to teach soldiers how to think and not what to think. Since operational art doesn't have to be predicated on doctrine, doctrines are not as important as the artistic views themselves (Strachan 2011a, p. 97). Doctrines and manuals are refined and finished products stripped of the creative process, the wild ideas, and innovative visions. They describe the established practices and procedures and not the philosophy of the art of war.

Three entire fields of warfare have been excluded and only occasionally mentioned. First one is space warfare because space should be de-weaponized and today or in the foreseeable future operational level commanders have no direct control of space assets (Warden 2000, p. xiv; Klein 2006; Odom 1993, p. 59). The two others are nuclear warfare and cyber warfare. The reasons for omitting them are two-fold. The first and more important reason is that in both the time-factor is almost eradicated. Art of war emphasizes nature of war as a reciprocal clash of wills. Cyber and nuclear wars could theoretically be initiated and fought in minutes and hours. There might not be time for the art in warfare. The second reason is that while all these three approaches to warfare have been thoroughly theorized they haven't been tested in practice. We have no proof whether dreaded escalation would occur or if nuclear powers could build enough restraints for the time-factor to remain meaningful. As Schelling (2008, p. 20–21) put it,

> This is a difference between nuclear weapons and bayonets. It is not in the number of people they can eventually kill but in the speed with which it can be done, in the centralization of decision, in the divorce of the war from political processes, and in computerized programs that threaten to take the war out of human hands once it begins. That nuclear weapons make it possible to compress the fury of global war into a few hours does not mean that they make it inevitable. We have still to ask whether that is the way a major nuclear war would be fought, or ought to be fought. Nevertheless, that the whole war might go off like one big string of firecrackers makes a critical difference between our conception of nuclear war and the world wars we have experienced.

This book follows the development of operational art throughout the ages along the theme of time, its control, and perceptions of temporality. As Wavell (1953, p. 33) advised younger soldiers, "when you study military history don't read outlines on strategy or the principles of war. Read biographies, memoirs, historical novels." War is an art form and thus a more literary method of research can be considered as suitable for the metahistory of the art of war. Thinking, planning, and conducting operations have an artistic element that turns mechanistic processes into operational art. According to Bülow (2013, p. 228),

> art is the application of science. Science is in the mind only; art descends from the mind into the sphere of activity. Art is all that can be done, whether good or bad: these qualities accord not with science; we know it, or we do not know it. We say of a science, that it is true or false; of an art, that it is good or bad.

This first chapter serves as an introduction to how warfare has been perceived over time. It divides the history of the art of war into three waves closely following the train of thought of Heidi and Alvin Toffler. It is argued that time and its meaning in art of war and operational art has been conceived in different eras in different ways as societies have progressed and character of conflict has followed suit. It argues that while war is an omnipresent part of human existence and its nature remains unaltered, over time the actual means and ways of warfare have been and will be in constant flux as time passes. Strategy as 'the art of the general' has become a politician's skill and operational art is the realm where soldiers exhibit their skills in time management.

1.2 The Three Waves as Temporally Divergent Ages of Warfare

> Because massive changes in society cannot occur without conflict, we believe the metaphor of history as 'waves' of change is more dynamic and revealing than talk about a transition to 'postmodernism.' Waves are dynamic. When waves crash in on one another, powerful crosscurrents are unleashed. When waves of history collide, whole civilizations clash. And that sheds light on much that otherwise seems senseless or random in today's world. (Toffler and Toffler 1995, p. 18)

As Colin S. Gray (2007, p. 3) stated, "wars are not free-floating events, sufficient unto themselves as objects for study and understanding. Instead, they are entirely the product of their contexts." To create a taxonomy of the historically omnipresent phenomenon of war it is helpful to divide the past into separate eras based on their cultural outlook and perception of temporality. There are perhaps as many categorizations as theorists. Hans Delbrück (1990a) divided the history of warfare into separate periods by tactical developments. These gave wars of their eras a common outlook. Fuller (1946) saw the history of warfare consists of six eras. These are the ages of valor, chivalry, gunpowder, steam, oil, and atomic energy. They are not rigid taxonomies but descriptions of the prominent characteristic of war during that period. Billy Mitchell (1999, p. 431) in turn wrote about the dawn of a specific "aeronautical era" that opened up with aviation technology after the "continental era" and the "era of the great navigators." His thinking focused on separating different eras of warfare from each other by the domains in which battles were fought.

For Alvin Toffler three dynamic waves of history and war created three distinctively different eras with clashes of civilizations occurring at when the waves collide. They are "*First Wave, or agrarian; Second Wave or Industrial; and now Third Wave armies*" (Toffler and Toffler 1995, p. 18). These waves are inseparably bound to perceptions of temporality thus providing a suitable framework for this book. Despite extensive criticism, Toffler's ideas were in the early 1990s influential in U.S. defense circles and profoundly shaped the U.S. military transformation (Lonsdale 2007, p. 232–233; Kagan 2006, p. 203–206). The first wave of civilizational and societal shift supposedly began around 8000 B.C. and prevailed globally until sometime around 1650–1750 A.D. when

> First Wave lost momentum as the Second Wave picked up steam. Industrial civilization, the product of this Second Wave, then dominated the planet in its turn until it, too, crested. This latest historical turning point arrived in the United States during the decade beginning about 1955—the decade that saw white collar and service workers outnumber blue-collar. (Toffler 1990a, p. 14)

Knowledge was the most important asset of this new era when knowledge workers surpassed the number of people laboring in factories. The Third Wave coincides with the advent of the Information Age. This new age began to emerge in the 1950s and has since kept accelerating its pace of evolution. Western theorists have written about

> a qualitatively different era to the study and practice of warfare. As a result, military periodicals and books have been replete with articles and essays on 'Information Age Warfare'. The problem is that the study of the military aspect of this advance in technology has been conspicuously lacking any sociological, philosophical, or theoretical component. (Leonhard 1998, p. 13)

Even if the First and to some degree the Second Wave are connected with development of mankind in general, the timing of the Second Wave implies reference primarily to the Western societies and the start of the Third Wave is determined by developments only in the U.S. The waves occur at different times in each society and when the Third Wave began to roll in the U.S. the undeveloped parts of the world hadn't even crested the first one. The waves cover long time-spans and become too all-encompassing to describe fine nuances of societal development. To understand evolution of the art of war, one must look at these waves in detail to spot emerging patterns.

As one wave starts to recede and the new wave comes crashing in, for a while the two waves interact and for the soldiers, generals, and the military systems they compose, strange and incalculable cross-currents emerge. One armed force is sucked in one direction while the other gets pushed in another. When the effects of waves cancel each other the art of war is said to become revolutionized. First wave civilization was a product of the agricultural revolution. The second wave civilization originated in the Renaissance but crested with the rise of industrialization, finally creating "what we came to call modernity—mass-industrial society, the civilization of the Second Wave" (Toffler and Toffler 1995, p. 18–19).

Different civilizations have fundamental differences in their conceptions of time. Traditionally in the U.S. everything has to happen 'right now, preferably yesterday.' This has only heightened during the Third Wave. In military affairs, as in most walks of life Russia with its agrarian past has had 'strategic patience' and the concept of time is not measured in hours, days or even months but rather by decades or generations (Adamsky 2010, p. 44). Waves have different characteristic and during the Third Wave "we are speeding toward a totally different structure of power that will create not a world cut in two but sharply divided into three contrasting and competing civilizations—the first still symbolized by the hoe; the second by the assembly line; and the third by the computer" (Toffler and Toffler 1995, p. 21). As Gray (2007, p. 249) noted, the closer one looks at the standard cut-off dates of periods, the less convincing dividing the past into neat eras and periods seems since fundamentally history and the passage of time are seamless. Using the three waves is beneficial when compared to the other stricter taxonomies, because the metaphor allows for the waves to overlap and interact.

The different waves are characterized by technological innovations that drastically altered the pace of warfare, allowing for more to be accomplished in a specific period of time than ever before. According to Toffler (1990b, p. 40.) "each culture has its own characteristic pace." This pace determines the speed of every activity of every society within that cultural or civilizational context—including warfare.

> The general pace of life, including everything from the speed of business transactions to the rhythms of political change, the pace of technological innovation, and other variables, is slowest in agrarian societies, somewhat faster in industrial societies, and races at electronic speeds in the countries transitioning to Third Wave economies. (Toffler and Toffler 1995, p. 247)

The different waves perceive the meanings of time, place, and force with their interrelations differently. Movement at the slow speed humans or animals could create characterized the agrarian age which had a relaxed attitude toward time as a commodity. For the early hunter-gatherers noting the passing of the present was sufficient, but for agricultural considerations time had to include a concept of the future as well (Stempel 2012, p. 39). Time became a question of when to do something to cause something else to occur in the future. In the agrarian age time was measured in passing of seasons, times for reap and harvest, and the rising and setting of the sun. An industrialized society could no longer adhere to this type of temporal conception. Industrialization created a whole new idea of time being money. Mass production required calculations how to produce as much as possible in as little time as possible. Hours, minutes, and seconds became the meaningful units of time and their measurement essential.

During the Second Wave in the Western world industrialization acted as the catalyst for mechanized warfare, but not all armed forces grasped the opportunity at once. Within each there was a competition between the progressive thinkers and the upholders of the traditional dogma. Old methods of warfare rarely become obsolete, but linger on in existence and the waves partially overlap. Many countries and especially intra-state actors of the Third World have had no other option than to stick to their agrarian or at best mechanized form of warfare. Like many so-called 'low-intensity conflicts' show, the old methods are surprisingly effective in countering new ones.

Occasionally wars themselves alter the course of future warfare and cause a paradigm shift from one wave to another to occur. This is not revolution per se, but a necessity to discard old practices and adopt new ones. When Germany in WW II attacked Poland the brave but doomed charges of the Polish cavalry symbolized an end of an era (Echevarria 2000, p. 216–217) just as well as the Battle of Shangani in Rhodesia where four British machine guns fought off 3000 attacking native warriors. In these occasions, we can argue that ways of warfare characteristic to two consecutive waves collided and resulted in a need to switch paradigms of warfare from one type of civilization to another. This in turn illustrated to the losing side a necessity to affect a societal shift as well. Fuller (1961, p. 98) used as an example the American Civil War that he called

> a war, not between two antagonistic political parties, but a struggle to the death between two societies, each championing a different civilization. Or, as Stephen Benét concisely depicts it:
>
> "The pastoral rebellion of the earth
> Against machines, against the Age of Steam."

Dividing the belligerent parties into the pastoral or agrarian South and the industrial lifestyle and machines of the North is a simplification, but descriptive. The speed of change and compression of time scale were different for both societies (Reid 2014, p. 7). The war was fought over the fundamental issue of who would determine the future of the U.S. (Bond 1998, p. 5–6); "would the rich new continent be ruled by farmers or industrializers, by the forces of the First Wave or the Second?" (Toffler 1990a, p. 23).

The American Civil War was still synchronized with the agricultural pace. This enabled generals to learn the art of war in their stride. As Sherman (1890, p. 220) wrote: "*I had to learn the tactics from books; but I was convinced that we had a long, hard war before us, and made up my mind to begin at the very beginning to prepare for it.*" Should a major war erupt today between two technologically advanced great powers there won't be time for commanders to learn their art by doing. American Civil War was the first great conflict that originated in the Industrial Revolution. It ushered in the industrial age and made the Confederacy abandon its pastoral past (Stempel 2012, p. 114–115). In many ways the American Civil War was an augur of the shape industrial war would take in the last part of the nineteenth and early twentieth century (Smith 2008, p. 84; Bond 1998, p. 68). Some have even called it the first modern war (Stempel 2012, p. 123–124). New technologies influenced the outlook of war and this was most evident in the tactical and operational levels while strategic thinking was not profoundly altered (Stoker 2010, p. 7).

The Second Wave, referred to here as "indust-reality," is divisible into two periods. The first one was characterized by steam as the motive power and railroads as the mobility-enhancing factor. The second period started when oil took over as propellant of movement and roads started to play a bigger role in channeling it. In both cases little attention was paid to the radical changes they created in the techniques of war (Fuller 1961, p. 311). Their direct effect on combat differed. When railways were used for transport, movement stopped at the rear of the battlefield. As Foch (1920, p. 350) put it, troops in the early industrial age "cannot be transferred from the left to the right of the battle-field. Modern extended fronts do not allow of this. One no longer has the time to affect it" The latter part of indust-real warfare can be called the age of motorization or mechanization since motorized movement had an impact within the battlefields.

WW II was the summit of indust-reality in the military sense. Many of the armies were initially at best motorized and could not brag about extensive mechanization. Yet mechanization allowed the full bloom of engine-produced movement on the battlefield. U.S. joined the war late, but its industrial production capability was unprecedented and by 1943 in tanks, heavy guns, and aircraft the U.S. production figures alone surpassed that of all the Axis powers combined. Thus, on one level, winning the war was a matter of grinding Axis strength down, but on another, it was question of maneuvering battles with new mechanized mobility enabled (Echevarria 2011, p. 145).

During indust-reality an integral part of the art of war was practical utilization of the principle of measured time. As industrialization hastened the pace of production it had an influence on first the society and then the way it wages its wars. When production was tied to the machines that needed no rest, time began to indeed equal money. Time grew and still keeps growing in importance at an accelerated pace since time is a scarce commodity in our societies. Indust-reality not only measured time anew but discarded the idea of cyclical time. It cut time into precise chunks and placed them "in a straight line that extended indefinitely back into the past and forward into the future. It made time linear" (Toffler 1990a, p. 104). Time gained a direction. This was a drastic change from the seasonal and cyclical understanding of time. For the industrialists an unused chance was gone forever since time was linear. For the agrarians a similar chance would come along sooner or later as part of a new cycle. The ones advocating cyclical notion of time generally have time at their hands. The industrialists favoring linear time wear watches on their wrists to measure it. There is no global universal rhythm. Societies and their armed forces are in different phases of development. Thus among the armed forces of the world the agrarian, the industrial, and the information society-related effects occur simultaneously and vie with each other.

The prevailing indust-real way of thought and style of living cannot be superseded in an instant, since it takes time to create a paradigm shift. Indust-reality became dominant after a struggle that lasted three centuries (Toffler 1990a, p. 289). We are experiencing a similar struggle in our minds and the surrounding reality. Not all ideas, structures, and systems will be replaced in a flash. Some persist in their existence longer than others, some adapt to the new conditions. We are living, once again, through a confusing historical period. Whatever will be the decisive course of development in the art of war, to understand how wars of the future will be fought, the adoption of modified principles and turning them into actual practices the military system requires "a cultural change, it cannot be achieved without widespread discussion, debate, experimentation, and ultimately, broad acceptance" (Alberts et al. 2000, p. 4).

This can be explained with the wave metaphor as well. As one wave breaks and the next one follows, there is a tumult when the advancing and receding waters cancel the effects of each other. "To understand today's colliding waves of change we must be able to identify clearly the parallel structures of all industrial nations—the hidden framework of Second Wave civilization" (Toffler 1990a, p. 25). We have to comprehend the essence of indust-reality and information age to determine which factors in human experience and war are of permanent nature and which are likely to change. We are living in between waves, waiting for the next one to sweep us along. We need to understand the characteristics of the currently dominant wave to ride the next.

> Again, to return to the argument of different civilizations advancing and developing at different paces, many of the countries earlier characterized as 'third world' comprise the emerging markets of today in which a huge technological boom is taking place. Hand in hand with the development of the industrial society the information society is simultaneously built. In our time pre-eminence in manufacturing has passed to the third world. (…) war between mass armies weighed down with baroque equipment they cannot use properly has become an established third world sport. This advanced world, too vulnerable to survive a war of attrition or mass destruction, must learn to conduct its affairs by the rapier—by the threat or use of small specialized forces exploiting high tempo and strategic surprise. (Simpkin 1985, p. 180)

Some countries are mentally and technologically well poised for Third Wave warfare but simultaneously have large industrial-age armies with corresponding weapons and doctrines. Third Wave societies have advantages of their own but they cannot compete in the realm of mass-produced weapons and equipment or agrarian methods of these 'multi-wave-societies.' They can potentially create enough indust-real mass that information age equipment doesn't suffice as a force multiplier to counter the mass. Perhaps an advanced Western information society must combine the rapier with the broadsword and even the flint ax to create operational art that can counter warfare characteristic to each of the waves.

What is the perception of the flow of time in the information age? It is, to say the least, full of discontinuities and phases when time moves at different speeds and is experienced in differently by individuals and societies. But what if time for our societies and civilizations was both linear and circular? This sounds like an inescapable paradox, but it is rather a question how progress is experienced. This we can chart by combining the metaphors of waves and cycles. Progress continues and common sense and history show us that "even if warfare moves in cycles, they are progressive cycles, and that each succeeding war in modern times between the Great Powers shows an advance mechanically on the last, and at least begins where the last left off" (Liddell Hart 1927, p. 55–56).

1.3 Waves and Cycles of Development in the Art of War

While waves represent the development phases of societies, Liddell Hart (1932, p. 273) argued that "military evolution seems condemned to travel in circles—to chase its own tail." Indeed, the development of paradigms in the art of war is of cyclical rather than on linear nature. Arden Bucholz (2001, p. 3–4) wrote about the cycles of war and in his opinion between 1864 and 2000 there have been three different cycles. Before 1860 warfare resembled the game of whist in which chance and ability to improvise made the winner. From 1864 to its fruition in the World Wars warfare was like bridge. The whole game plan had to be ready when the game commenced and followed a preprogrammed pattern. In the twenty-first century warfare is like go. The characteristics of relatively uniform pieces depend on their positioning in the overall mosaic. Once the pattern is set, victory or defeat are decided. The paradigms of war are constantly changing but Toffler's wave metaphor is so suitable that we shall further extrapolate how the waves interact with the cyclical development patterns the military abides to. Waves have a rhythm of their own just as warfare does. In its traditional forms warfare

> permitted maneuver and countermaneuver, attack and counterattack, and movement and pause. It also gave rise to the phenomenon known as the culminating point in campaigns, that point at which the campaign is in near equilibrium where the right effort on either side can have significant effect. All of our thinking on war is based on serial effects, on ebb and flow. (Warden 1995, p. 19–20)

The same wave-pattern applies to development of warfare and operations within a war. Using the wave metaphor "helps us see beneath the raging surface of change. When we apply the wave metaphor, much that was confusing becomes clear" (Toffler 1990a, p. 5). The receding and incoming waves create temporary confusion during which the direction of movement is unclear and undecided. It is worthwhile to combine the wave metaphor of Occidental thought with the oriental cyclical concept of time. If, instead of treating time as teleological, we argue that whatever goes around comes around we can conceive of development as a cyclical process.

> There appear to be alternative and plural 'times' operating under different rules in different parts of the universe or universes we inhabit. All of which knocks the props from under the Second Wave idea of universal linear time—without substituting ancient notions of cyclical time. (Toffler 1990a, p. 297)

The wave metaphor can be employed to understand the characteristics of the present. The point is to prepare for and to shape the future at the same time. The premise is that nothing remains unchanged and the future is always fluid instead of frozen and predetermined. The future can be changed daily by every decision we make (Toffler 1990a, p. 129).

Military evolution follows the pattern and direction of societal development but generally lags behind—by several decades in the worst situations. A good example of what this means to the military is visible in the influence of indust-real pace of business. A crisis of leadership results from and is intensified by the accelerated speed of events and need for decisions. More and more decisions have to be made faster and faster and response times are minimized (Toffler 1990a, p. 229). The armed forces are attempting to adapt to both the acceleration in the speed of change and the speed of decision-making alike. John Boyd's OODA loop composed of four successive phases of Observe, Orient, Decide, and Act started to develop from a rudimentary attempt to get fighter pilots to make their combat decisions quicker to cover situations in which a system composed of multiple individuals could be streamlined and taught to make better decisions faster. Thus the OODA loop is not such a simple concept as it often is mistaken for. Decision-making cycle rather consists of numerous interconnected cycles, and the idea of getting inside the enemy decision-making cycle requires more than just being faster than him. It means also getting inside his mind. Somewhat paradoxically once the OODA loop starts to spin, it must be forced to accelerate. Success is the biggest trap in the OODA loop, since it may cause action to stop (Coram 2002, p. 336–339). Cycles and waves as concepts are interconnected and there are two options to choose from: either to optimize adherence to these patterns or to attempt to break away from them and increase preexisting discord in the rhythm one's potential adversaries employ.

Perhaps the most suitable means can be created when one combines the lessons of military history to the characteristics of new temporal and national context. Civil progress reshapes the character of war since they "develop from out of the central idea of each cultural cycle" (Fuller 1946,

p. 27). Nevertheless, culture is a social construct that changes constantly, albeit slowly. It is always in a flux and assigning too specific characterizations to short periods of time would be at odds against the idea of culture itself. Society produces its culture as well as its warfare and we should concern ourselves with the megatrends of change. However,

> as war is changed by civil progress, so also is civil progress changed by war—there is a reciprocal action between them. Further, that war is the one permanent factor in its changings, for whether the period under examination be predominantly religious, commercial or industrial, and whatever its political and social systems may be, war is never absent. [...] though military systems also change, war is never annihilated. Except for brief periods of lassitude, the evolution of the weapons and means of war has been continuous and progressive. (Fuller 1946, p. 26–27)

It occasionally seems, especially when viewed in the light of such organized mass slaughters as Somme and Passchendaele, that the art of war had not undergone evolution but rather devolution during the high-tide of the Second Wave. This occasionally happens in the history of warfare. Some weapon gains such dominance for a certain period of time, that existing tactics are useless against it. Then and there, tactics undergo a short period of decadence. The invention and increasing use of machine guns created such temporary confusion in tactics. In the words of von der Goltz (1906, p. 175), "modern battles are decided by great masses of projectiles simultaneously hurled at the enemy." The effect was evidenced in cavalry tactics, which had to be rethought. While horses disappeared from battlefields in one phase of the cycle, a new type of cavalry returned in concurrent phase. Liddell Hart (1927, p. 50) dissected war into three basic elements; those of guarding, moving, and hitting. The initial value of cavalry had been its mobility, and once cavalry disappeared, warfare became stagnant. The advent of the tank as indust-real cavalry promised a mobile future when moving and hitting would be enabled again by the protection armor offered. Development of doctrines and tools of warfare move in cycles in which the preponderance of attacking and defending alternately dominate. The ways and means of warfare adapt to changed situations and the cycle revolves. After the disastrous, static, and bloody WW I tactics took huge leaps forward in the time leading to WW II. Mass production and mass armies—just as well as mass destruction—had to be replaced with a new way of thinking that would revitalize mobility.

It is possible to retrospectively emplot the past and impose wave or cycle patterns on it, but the future is always fluid. The more daringly military thinkers have tried to envision the future, the more spectacular have been their failures. Nevertheless, if they didn't attempt to predict the outlook of a future war, there would not be blueprint for development. Even the most perceptive predictions coming true in detail "is unlikely because military prophecies are nearly always wrong, but it is essential to consider future developments or else there can be no general guiding policy for the present" (Martel 1931, p. 118). While superpowers can use considerable amounts of money on research and development of new technologies, small states are seldom offered such luxury.

The idea of cycle fits the development of war-fighting tools, theories, and practices well, but adopting the concept of a wave allows us to understand the periods in between the cycles as riptides when ideas mature. Development never grinds to a standstill. New ideas emerge constantly but many of them remain unrecognized. Some ideas emerge at a profitable time and ride the crest of the wave, some drown in tumultuous waters between the consecutive waves. Some ideas actually initiate a cycle that amplifies the wave itself and makes it prevail longer than its original momentum would have allowed for. This is what railroads, combustion engines, tanks, aviation, and then the invention of the rotary wing in turn did for the Second Wave of indust-reality.

> Americans, as a race, are the foremost mechanics in the world. America, as a nation, has the greatest ability for mass production of machines. It therefore behooves us to devise methods of war which exploit our inherent superiority. We must fight the war by machines on the ground, and in the air, to the maximum of our ability. (Patton 1947, p. 366)

The wave continued by picking up extra momentum. In the long run the frequency of waves might increase, since if everything in society has become faster and faster as a byproduct of civilizational development, it is entirely plausible and even predictable, that the development cycle will pick up speed as well. 'Revolutions' are likely to follow each other in shorter succession than ever before. National strategies tend to look at least ten years ahead, and one of the reasons is that by choosing such a time-span, the strategy-cycle is synchronized with the average procurement cycle of most defense equipment (Strachan 2013, p. 235). As long as the cycle of civilian technology development accelerates, there is a pressing need to get the military development cycle to emulate that speed in development of strategy, concepts, and doctrine. This has regrettably seldom occurred. As Leonhard (1998, p. 7) wrote, “doctrine typically lags behind technology. Sometimes leading to disaster.” Alberts et al. (2000, p. 201) summarize this requirement of having the cycles revolve at the same speed by saying; “what is needed is an approach that synchronizes the development of military strategy and doctrine with the advances in technology and with the technology insertion process.” It is always a question of synchronization. The speeds of the cycles don’t need to match as long as their phases are synchronized so that an output from the civilian technology cycle influences the doctrine and concept development cycle at the right time as input.

We can say that the change in the organization, tactics, and weapons of an army has developed more or less at the same speed as the conditions of war have changed and this, in turn, has occurred following the pace of technological development of the society. The problem is that each of the above has occurred at intervals. This has led to a situation in which “the armies of today are as helpless, and their prospects as hopeless, as a portly policeman trying to catch a motor thief—the thief of time” (Liddell Hart 1932, p. 185).

1.4 Cresting the First Wave and the Civilized Wars

> As the weapons of war change, so does the character of war change, and though this is an undoubted fact, tactically it must not be overlooked that weapons change because civilization changes; they do not change on their own account. To-day wars arise out of economic causes, because our present civilization is an economic one, its master pivot being the machine in one form or another. As the present age is largely a mechanical one, so will the wars of this age take on a similar complexion, because military organization follows civil organization. (Fuller 1943, p. 9)

Fuller meant that the ways and means of warfare are mirror images of the functions and characteristics of the society or civilization that wages them (Matheny 2012, p. 3–7). The reasons for war as well as the character of warfare can be derived from the outlook, composition, and development phase of the society. Great Britain at the peak of its international dominance was a maritime society, and thus its military capability mostly consisted of an unconquerable navy. The society of the Middle Ages was agrarian and its elites consisted of landed gentry. The battles were often fought between a peasant mob armed with scythes or pitchforks and the knights in chivalric armor and weaponry. The Industrial Revolution enabled mass production of weapons and their rapid transportation with the aid of first steam and then internal combustion engine. It is seldom that military progress sets the pace for the society. However, there will never be a war that doesn't reflect the stage of development of its societal and cultural structures but sometimes war itself changes the society and its structures (Shy 1976, p. x).

War has been with us for a long time and as a byproduct of civilization is likely to remain with us in the future as well. It first emerged about 12,000 years ago when early civilization was born as humans began to coalesce into small towns and villages in Mesopotamia (Stempel 2012, p. 6). As civilization progressed, so did the methods of fighting. "War to the men of the Stone Age was not the business of a selected few, it was the occupation of every adult male, and it is still so, with the addition of numerous women. In savage warfare, the aim was to kill all enemy males and abduct the women and children" (Fuller 1961, p. 40). In primitive societies often the entire tribe engaged in war. As the hunter-gatherer societies of nomads moving their adobe and their families with the seasons developed into agriculture-based societies, a division of labor began among the population. An agricultural community as a proto-society could rudimentarily organize its warlike endeavors. There were those who farmed the land securing nutrition and those who defended the crops and the stable adobes of the society. The society was divided into warriors and others they protected (Fuller 1961, p. 31). War proper emerged when first agriculture and then early stages of politics were created (Stempel 2012, p. 35–39).

This led to a gradual distancing of non-combatants from the war. Although in many cases wars ravaged the land and destroyed civilian life and property, they were not the main targets and as centuries passed, the civilians were increasingly protected. The development story of warfare shows us how at different periods of history wars themselves "came to be modified, and gradually humanised. It is a story of 'ups and downs'—but far more up than down" (Liddell Hart 1950, p. 367). Our society today could not withstand the approach of Genghis Khan who often ordered everyone including the domestic animals to be killed so that no living creature should dwell in what he conquered (Juvaini 1958, p. 133). The old-fashioned berserker warrior has no place in a stable society. As people fled the countryside to follow the siren song of the cities, the warrior was no longer a peasant and the way he fought changed. Industrial Revolution began an accelerated growth of technology in all spheres of life and the warrior of the time mirrored the societal conditions. As Charles De Gaulle (1976, p. 53) wrote,

> modern conditions of military action demand, therefore, constantly increasing technical skill from fighting men. The equipment, which the force of events has introduced into the ranks, demands the gift, the taste, the habit of serving it. This is a consequence of evolution, ineluctable in the same way as the disappearance of candles or the end of sundials.

Technology has throughout history had a tremendous impact of the outlook of wars. A period of accelerated development of technology began with the advent of industrialization. But has the heart and soul of war changed at the same pace the technical tools to wage it have? Definitely not. Mary Kaldor's argument about "new wars" is questionable, but she was right in claiming that the prevalent conception of nature and outlook of war is a Western phenomenon.

> What we tend to perceive as war, what policy-makers and military leaders define as war, is, in fact, a specific phenomenon which took shape in Europe somewhere between the fifteenth and eighteenth centuries, although it has passed through several different phases since then. (Kaldor 2007, p. 15)

In the twentieth century the West was technologically most advanced and led development of warfare. The attempt to be faster and faster has been an Occidental tendency, but oriental thought has influenced Western operational art and strategy profoundly throughout the ages. A necessary starting point to treat the art of war philosophically is to accept war as a universal part of the human condition. It is neither desirable nor avoidable. Oriental thinkers of the past accepted this and for example Ssu-ma wrote "Thus even though a state may be vast, those who love warfare will inevitably perish. Even though calm may prevail under Heaven, those who forget warfare will certainly be endangered" (Ssu-ma 1993, p. 126). Politically the love of war is just as dangerous as turning a blind eye to its possibility. For the Chinese thinkers war was not an end of policy but an evil rendered necessary by an imperfect world (Creveld 2005, p. 29).

The role of war is central in history. As Christopher Coker (2010, p. 50) argues, war is crucial to the story of mankind. This doesn't mean that war should be glorified. "War is the most terrible fate that mankind can envision. Men must bravely face it when it comes on us with iron necessity, but at the same time recognize that it is a crime to encourage it unnecessarily" (Delbrück 1997, p. 99). War seems to be a recurring event that directs progress. At times war makes us evolve in bounds, at other times it takes civilization back. The ultimate form of regression would be an all-out thermonuclear war. War gives history its direction and structure and helps to set other events into causal relationship. "*War, in a word, is the medium of history*" (Coker 2010, p. 59).

Fighting is societal activity and if wars are to be curtailed, only civilization can accomplish this feat. Helmuth Karl Bernhard Graf von Moltke claimed that the answer to limiting the totality and destructivity of war will not come from practitioners of the art of war. His military genius was universally accepted, but simultaneously he was occasionally seen as too 'soft' and some claimed that he was more disposed toward peace than war (Ludendorff 1919, p. 56). Depending on the viewpoint this can be read as highest possible praise. Even if at heart he was a man of peace, as a practitioner even he could not change the nature of war to be less devastating (Gat 2001, p. 326–327; Bucholz 2001, p. 10).

> Whoever knows war will agree that it cannot be restrained by theoretical chains. Lessening its horrors is rather to be expected from the gradual advance in general civilization that promote the humanity of each individual. This is because the conduct of war reflects the progress of civilization. Only such general progress, and not laws of war, can lead to the goal. (Moltke 1993, p. 23)

Civilization can make wars less destructive. When the value of human life is acknowledged, measures are taken to limit wanton destruction. Even if the 'new wars' or intrastate wars of the post-Cold War era hold strong elements of genocide, they are no comparison to the wholesale slaughters Julius Caesar (2010, p. 43–48) organized in Gaul. Liddell Hart (1950, p. 366) argued that Hitler seemed a gentle man in comparison to this "much praised missionary of Roman civilization." Warfare in the past was more violent, and varying restrictions on its conduct have been applied at least since the Middle Ages. There have been and there will be temporary setbacks, but history shows that wars become less destructive as civilization progresses and limits warfare in many ways. After all, one goes to war to win the peace, and unrestricted violence would make the war counterproductive (Schelling 2008, p. 126).

Caesar wasn't unnecessarily barbaric in applying his art of war. As Fuller (1965, p. 182) summarized his essence as a commander-in-chief, "he fitted his means to his end: he was neither a devil nor an angel, he was a craftsman." Caesar was clear-sighted enough to understand what was required and was unscrupulous enough to execute it. He was an exemplary specimen of the pinnacle of development in warfare and civilization of his time. In the words of Clausewitz (1989, p. 101), "Possession of military genius coincides with the higher degrees of civilization: the most highly developed societies produce the most brilliant soldiers, as the Romans and the French have shown us." Caesar was a brilliant commander-in-chief, but even the Roman civilization, representing the best in the Occidental world, didn't require limiting the suffering inflicted on barbarian enemies. When he had to fight Roman legions and the enemy consisted of his own countrymen, Caesar chose a different approach. In a letter he described his chosen method as almost a predecessor to what the U.S. attempted in Vietnam.

> Let us see if by moderation we can win all hearts and secure a lasting victory, since by cruelty others have been unable to escape from hatred and to maintain their victory for any length of time (…) This is a new way of conquering, to strengthen one's position by kindness and generosity. (Caesar, cited in Fuller 1965, p. 182)

Had he chosen to treat fellow Romans as enemies in the same cruel manner he employed against the barbarians, the result would have been a Pyrrhic victory. Caesar would have emerged victorious, but the desire for revenge would have discredited possibilities for a prolonged peace. He understood that such a devastating war as he waged on Gaul would not have worked in Italy. In a civil war one has to win the goodwill of population and not only subdue the adversary (Fuller 1965, p. 181).

In study of operational art it is unnecessary to evaluate whether someone was a good person but focus on his professional abilities as a commander (Wylie 2014, p. 15–16). Many great captains of the past and present are not among the great humanitarians of all time. They were artists, but their masterpieces were painted with blood. When it comes to the art of war, only their leadership and command matter. Napoleon was an egotist, Cromwell a zealot, Sherman devoid of compassion, and Suvorov a misanthrope, but they performed their military tasks admirably (Jomini 2007, p. 15–16; Delbrück 1990d, p. 188–191; Sherman 1957, p. 11; Longworth 1965, p. 132). Some combined the good with the bad. Often they were opportunistic and ruthless men to whom the ends justified the means and this guided their operational art. In discussing Caesar's methods of fighting first the Gallic and then the civil wars, Fuller (1965, p. 55) wrote;

> In the one, the atrocities he perpetrated on the unfortunate Gauls have seldom been exceeded by a civilized soldier; in the other, his leniency toward his enemies has seldom been equaled even in more recent civil wars. This again reveals his amorality. He was a man totally governed by his end, and whether the means he employed to gain it were good or evil meant exactly nothing to him.

The Roman Empire didn't fall victim to a more powerful contender but imploded. Barbarians never conquered Rome, but they were accepted into the empire. The Germanic barbarians in time came to replace the Roman soldiers in the legions (Delbrück 1990b, p. 220). The collapse of the Roman Empire was not caused by the barbarians corrupting the "purity" of the empire but by assimilating too well into it. It took over two centuries, but the barbaric Germanic and Gothic tribes who first took over the protective duties of the empire became civilized and "in the atmosphere of civilization, their warlike nature melted away along with their barbarism" (Delbrück 1990b, p. 419; 1990c, p. 66). Rome didn't overextend itself nor did a rival defeat it. It simply became too civilized to hold itself together. Civilization creates culture and culture is unfavorable to the art of war (Gat 2001, p. 355–356). Traditionally civilization was always under threat from the barbarians occupying its peripheries and had to defend its borders actively (Keegan 1998, p. xiii). This applied to Rome against the German tribes as well as China against the Mongolian hordes. During the Third Wave the situation is different. As Bertrand Russell (1959, p. 83) wrote, "It is not now barbarians who constitute the danger. On the contrary, it is those who are in the forefront of civilization."

From the Peloponnesian War onward mercenary soldiers dominated warfare for a long time. Knights were the most expensive mercenaries and simultaneously the most unreliable. Once a battle seemed lost, the knights would dash off on their horses and leave the masses of foot soldiers to die

(Delbrück 1990c, p. 455). The light mercenary infantry troops were relatively cheap to raise and in time their efficiency increased (Fuller 1960, p. 42). Professional soldiers remained through the Middle Ages the primary tools in the art of war. Swiss mercenaries were highly valued, and Switzerland was perhaps the only place where the art of war progressed during the Middle Ages (Delbrück 1990c, p. 429). For the Italian *condottieri* war was a business and mercenaries had no desire to lose their lives in the service of their employer while the princes employing them didn't wish to incur any risk. "The soldier of our day is a merchant. So much of my flesh, of my blood, is worth so much. So much of my time, of my affections, etc. It is a noble trade, however, perhaps because man's blood is noble merchandise, the finest that can be dealt with" (du Picq 1987, p. 244). Warfare was relatively humanitarian, since soldiers didn't wish to waste their lives, and risks taken were compensated for with money. Money could buy soldiers to fight for the belligerents who kept a distance between themselves and the whole messy business. Mercenaries were a necessity, since without them, "rich man without arms must be a prey to a poor soldier" (Machiavelli 1965, p. 204).

Machiavelli discredited the mercenaries and wanted the army to consist of citizens. He disliked the *condottieri* because their armies were composed of horsemen and he considered infantry to be the decisive arms just as it had been for the Romans whom he idolized (Delbrück 1990d, p. 101–113; Creveld 2008, p. 96–97). The turning point back to the concept of citizen-soldiers occurred in Napoleonian warfare and subsequent rise of the mass armies. Instead of being the profession of relatively few warriors, war again became everybody's business like in the primitive societies. Napoleon's novel thinking on war was the starting point of creation of mass armies through mobilization and was transformed later into the concepts of total mobilization and as a result to paradigms of total war (Heuser 2010, p. 137–138).

Liddell Hart argued (1950, p. 369) that the eighteenth century was a turning point in the customs of war, and the ability to reduce its evils was a great civilizational achievement that opened up a prospect of progressive limitation of war. For Bond (1998, p. 14–15) warfare of that time was very limited in comparison to the religious wars that preceded it and the revolutionary and nationalist wars that followed it. Whenever there are setbacks in the tendency of wars to become less destructive, these are caused by either political or technical developments that manage to destabilize the existing order (Liddell Hart 1950, p. 370). When social forms and the stability of status quo are shaken, war regains some of its savagery. We saw this happen after the collapse of the Soviet Union and in the War on Terror. As Coker (2010, p. 11) noted, "War is protean—it adapts to the external environment: it changes culture by culture, and over time."

To return to the American Civil War, an often-quoted comment from Robert E. Lee tells everything about his attitude. Lee stated during the Battle of Fredericksburg that "it is well that war is so terrible, otherwise we should grow too fond of it." To save the Confederacy from more casualties Lee surrender his troops when he deemed the ultimate loss inevitable. He was a man infused with the Southern agrarian traditions. Opposing Lee were men of different breed. As Fuller (1961, p. 107) described them,

> Sherman, and to a lesser extent Grant, Sheridan, and other Federal generals belonged to the age of the Industrial Revolution, and their guiding principle was that of the machine which was fashioning them—namely, efficiency. And because efficiency is governed by a single law, that every means is justified by the end, no moral of spiritual conception, or traditional behavior, can be tolerated should it stand in its way. Sherman was the leading exponent of this return to barbarism. He broke away from the conventions of nineteenth century warfare, and waged war with steel as ruthlessly as Calvin had waged it with the world.

Sherman favored total war that he waged on the civilian population of the South as well as its army. He felt no remorse, stating bluntly "if the people raise a howl against my barbarity and cruelty, I will answer that war is war (…) If they want peace they and their relatives must stop the war" (Sherman 1957, p. 11). It is an accomplishment that in such a violent art form as war one man can have such an impact that the destruction he caused is still resented a century and a half later (Davis 2001, p. 13). Yet Sherman's march to the sea was not senseless rampage, but a series of deliberate acts stemming from the strategic need to destroy the material base of the South (Smith 2008, p. 88). There is no reason why the technically more advanced 'civilization' should be that ethically. Warfare can't be isolated from its cultural, political, social, and temporal contexts. They greatly influence the general character of war (Bond 1998, p. 15–16). Politics and policy can

> reduce war to something tame and half-hearted. War is often nothing more that armed neutrality, a threatening attitude meant to support negotiations, a mild attempt to gain some small advantage before sitting back and letting

> matters take their course, or a disagreeable obligation imposed by an alliance, to be discharged with as little effort as possible. (Clausewitz 1989, p. 218)

We saw during WW II that nations declared war on each other based on alliances without their troops ever getting engaged in actual fighting. War has political purposes. Sometimes it is not waged but employed as gradually increased and rationally calculated pressure.

> War is simply a continuation of political intercourse, with the addition of other means. We deliberately use the phrase "with the addition of other means" because we also want to make it clear that war in itself does not suspend political intercourse or change it into something entirely different. In essentials that intercourse continues, irrespective of the means it employs. (Clausewitz 1989, p. 605)

When war is viewed as a political tool, there are many ways of employing it and absolute war is only the penultimate end of the spectrum (Bellamy 2007, p. 17–19). War continues political dialogue with extra emphasis. It can be used gradually with new elements included, should the fulfillment of political goals demand extra pressure. Intercourse continues on all levels simultaneously and warlike measures are added to coerce the enemy. "In short, at the highest level the art of war turns into policy—but a policy conducted by fighting battles rather than by sending diplomatic notes" (Clausewitz 1989, p. 607). Traditional diplomacy and other forms of political communication never cease.

Diplomacy of the past was different from that of today and so is warfare. War has indeed become "not so much a contest of military strength as a bargaining process—dirty, extortionate, and often quite reluctant bargaining on one side or both—nevertheless a bargaining process" (Schelling 2008, p. 6). Warfare reflects the characteristics of belligerents because all wars were products of the societies that fought them (Howard 1983, p. 48).

> War is a product of its age. The tools and tactics of how we fight have always evolved along with technology. We are poised to continue this trend. Warfare in the Information Age will inevitably embody the characteristics that distinguish this age from previous ones. These characteristics affect the capabilities that are brought to battle as well as the nature of the environment in which conflicts occur. (Alberts et al. 2000, p. 1)

Naturally at any given age the most sophisticated means, be they longbows, muskets, Spitfires, or autonomous systems, are employed in warfare. We live in an age of abundant information and this affects our methods of war but also our minds and the way we live our lives. Wars are fought to secure a better peace and not maximize destruction. Still, violence is an integral part of war and cannot be completely eliminated. Smith has claimed that warfare is brutal because

> force is applied by military forces armed with lethal weapons. When unleashed, they will kill and destroy. That is what they are trained to do—and that is actually what we, civil society, ask of them. However, this is an unspoken contract, which is encased within the clear frameworks of war and peace that have evolved over the ages, but most especially in the past two centuries. And the fact that these no longer suit the reality in which we live does not stop us from rearranging reality into the frameworks we know. (Smith 2008, p. 27–28)

A persistent part of our contemporary thought seems to be incredulity toward the possibility of a major war in the future. Earlier some military thinkers claimed that a perpetual peace would not even be sustainable for mankind since it could only be purchased at the price of man's nobles qualities and destinies (Foertsch 1939, p. 7; Freytag-Loringhoven 1991, p. 347). This may be a step too far, but wars have pushed societies and individuals to develop themselves. "Permanent peace is a dream and not even a beautiful one, and war is a law of God's order in the world, by which the noblest virtues of man, courage and self-denial, loyalty and self-sacrifice, even to the point of death, are developed" (Moltke, cited in Gat 2001, p. 327). Wars have occurred for all of recorded history and they are despite man's noblest aspirations not likely to ever vanish completely. The old proverb that history always repeats itself has a grain of truth, but only because statesmen and commanders have paid insufficient attention to the history of war and records of past conflicts. Therefore, while history doesn't repeat itself, "it is the ignorance of history which does so, with the result that identical blunders recur in every age" (Fuller 1960, p. 306). Ignorance instead of belligerence is the most dire threat to peace. Wars are fought for no gain time after time.

Military historians extol the way Polish economist Jean M. Bloch predicted in the first years of the twentieth century the characteristics of a future World War. He claimed that war "will become a kind of stalemate, in which neither army being able to get at the other, both armies will be maintained in opposition to each other, threatening each other, but never being able to deliver a final and decisive attack" (Bloch 1914, p. xvi). He foresaw the immobile nature of the war to come, arguing that it would "be a great war of entrenchment. [...] Battles will last for days and at the end it is very doubtful whether any decisive victory can be gained" (Bloch 1914, p. xxvii). There are many perspicacious arguments in his hefty book, but ultimately all of them are nullified by the main conclusion. Bloch made precise calculations of the resources of European powers to argue that

> war has become impossible alike from a military, economic, and political point of view. The very development that has taken place in the mechanism of war has rendered war an impracticable operation. [...] Thus, the great war cannot be made, and any attempt to make it would result in suicide. (Bloch 1914, p. xvi)

Such an outright denial of a possibility of a large-scale war reduces the importance of his predictive illustrations. Bloch described what he thought would never occur and by portraying future wars in such a destructive way he wished to underline their irrationality and impossibility. Sherman had a more realistic view of the economics of war since he wrote in a letter that

> the cost of the war is, of course, to be considered, but finances will adjust themselves to the actual state of affairs; and, even if we would, we could not change the cost. Indeed, the larger the cost now, the less will it be in the end; for the end must be attained somehow, regardless of loss of life and treasure, and is merely a question of time. (Sherman 1890, p. 370)

Wars will be fought even if logic dictates that belligerents cannot afford them. Economic suffering never led to shorter wars. Wars are fought for material gain, but unfortunately this gain can be of almost any type, since for some for example symbols and beliefs are just as material as land or resources (Dillon-Reid 2009, p. 15). Even unprofitable wars possess unquestionable ability to fascinate men (Creveld 2008, p. 1). Furthermore, from the economic perspective logic would dictate that it would be beneficial to sacrifice money and resources heavily in the early stages of the war to prevent the conflict becoming prolonged and draining the state. Time is money, since the longer the conflict, the higher the accumulated costs will be in the end.

It speaks volumes of the inseparable bond between the human condition and warfare that not one but two World Wars broke out in the three decades following the publication of Bloch's book. There was a flawed assumption that rationality would prevail and short-term conditions would seem overriding (Freedman 2006, p. 6). Bloch's expertise consisted of taking a truly multidisciplinary approach into understanding warfare and especially the economic, political, and social drivers that could lead to war (Echevarria 2000, p. 221; Mandeles 2005, p. 52–53). Despite his careful calculations and reasoning Bloch was proven wrong. He lamented how "in preparations against that impossible war that these so-called practical men, who are the real Utopians of our time, are wasting the resources of civilisation." (Bloch 1914, p. x). Bloch as a pacifist was ultimately who turned out to be the Utopian optimist (Bond 1998, p. 88–91). A more realistic prediction was that

> though the principles of modern warfare may demand the most rapid decisions, and though, perhaps, those principles may lead to bloody battles at the very beginning of the struggle, it is yet probable that the whole result will take the form of a severe contest, in which the combatting armies, as followed on the map, either move but little from the spot, or, in comparison to the extent of ground involved, make but very insignificant progress. (Goltz 1906, p. 89)

The bloodiest battles in WW I took place after armies had become immobilized in their trenches. This was mostly due to utter disregard of commanders on both sides for the lives of soldiers. The importance of swathes of land was grossly overestimated and thousands of lives were wasted fighting over a few meters of soil. Generals "often gave away in stupidity what they had gained in ignorance" (Lawrence 1997, p. 210).

One of the most visible characteristics of twenty-first century warfare so far has been the role of high technology and one of the least visible the decision to withhold the full capability of high-technology weapons (Strachan 2013, p. 167). Based on self-imposed restrictions to the destructive potential we can claim that the Western art of war has become more 'civilized,' but there are other indicators as well. The unwillingness to commit troops and endanger them more than is absolutely necessary has created fiascos like Afghanistan. War is deadly business. Freytag-Loringhoven (1991, p. 206) extolled the professionalism in Ney,

> which enabled him to say to a wounded soldier who begged for help during the retreat from Russia in 1812, "How can I help you? You are a part of the wreckage of war." In these quietly spoken words there was no cruelty; there was only the expression of an old truth—the soldier in war is doomed to destruction; a truth almost forgotten in our peace-loving age.

In confronting attrition tactics and strategies Third Wave societies would be extremely vulnerable to pressure from within to end the war. U.S. and other Western armed forces have created operational plans and even entire strategies that enable use of force without suffering casualties of one's own. "We fight so as not to lose the force, rather than fighting by using the force at any cost to achieve the aim" (Smith 2008, p. 19). This type of warfare has been labeled "virtuous" by James Der Derian (2001, p. xiv–xv) or "Post-Heroic" by Edward Luttwak (1995; Kaldor 2007, p. 29). In warfare the excessive civilized element has led us from heroic close combat to striking from distance. Mitchell (1999, p. 443) hailed air power as the "*greatest civilizing element in the future*" because it could reach anywhere in the world, and beyond exerting power, it could be a force for good. Air power has been efficient, but its civilizing mission has failed. At the core of virtuous war lies the attempt to minimize casualties and use technical capability to threaten and to actualize violence from a distance. While it attempts to distance war and the dying soldiers from civilians, "*in the final analysis that it seeks to evade, virtuous war is still about killing others*" (Der Derian 2001, p. xv–xvii). Leonhard saw this tendency of non-committance as a fallacy of U.S. operational art.

> American military decision makers as a whole believe that the great goal of warfare is to blow up the bad guys at extreme distances. Artillery or air-delivered fires can pretty much solve any military challenge to our nation, if we commit enough funding to indirect-fire weapons, their munitions, and the capability to detect the enemy. The argument in favor of long-range fires is compelling, well respected, universally agreed to, and flat wrong. Belief in the effectiveness of indirect fires is the single most commonly held delusion in American history. (Leonhard 1998, p. 210)

Post-Cold War history has shown that air power without ground troops is unlikely to lead into conclusive victory. Post-heroic societies are not eager to go to war, but momentarily may be fired with enthusiasm by constant mass media excitement (Münckler 2007, p. 228). In operational art, heroic and post-heroic are just labels attached on different ways of fighting. Post-heroic generally refers to the fact that damage on the enemy is inflicted from afar with little or no personal risk. In the antiquity when the Persian and Greek armies dominated warfare, fighting centered on heroes and their deeds. Shooting from afar was cowardly and didn't constitute a proper test of manhood since someone like Paris was able to kill the great Achilles (Creveld 1991, p. 80–83). "War was still in its heroic phase; much of the decisive fighting took the form of duels between heroes. Leadership was personal and not delegated: a general-in-chief led his army into battle and did not direct it from the rear" (Fuller 1960, p. 152). Coker (2004) is among the many who have lamented the end of the heroic phase in warfare and subsequent demise of military heroes and for all of his Information Age ideology Leonhard (1998, p. 216) viewed warfare as great heroic drama that "will expose man's meanest nature and worst characteristics: greed, ambition, indecision, cowardice, cruelty and hatred. It will also provide a stage upon which man can give expression to his virtues: selflessness, loyalty, courage, restraint, and love." Yet, perhaps, life is qualitatively better without Achilles, Hector, or other Homeric heroes among us. Maybe moving into post-heroic period in combination with the ever-increasing destructive potential of warfare will help curtail war. Unfortunately the same impulses that lead to minimizing casualties through stand-off precision strikes in post-heroic forms of warfare simultaneously place restrictions of operational art and diminish the chances of success (Lonsdale 2007, p. 241).

1.5 Politics, Strategy and the Art of War Over Time

> War at all times, whether civil war between sections of a common country or between nations, ought to be avoided, if possible with honor. (Grant 2004, p. 241)

The goals of politics and military strategy differ fundamentally. As Ulysses S. Grant noted, war must be avoided and that is the task of politicians. Of course strategy should not drive a nation into war either, but if a war erupts, the military has to support policy by winning the war or at least creating a situation in which peace may be restored. The age-old accusation of armies preparing for the last war may be inaccurate, but Smith concedes that they often prepare for the wrong one, because the government only funds preparations against the primary enemy (Smith 2008, p. x).

Strategy struggles to cope with the political demands and balance them with the resources allocated. One way to approach the troubled marriage of politics and strategy is to view the affairs of the state through the prism of grand strategy. Alexander can be used as an example how strategy was consistently subordinated to policy and this is the essence of grand strategy (Fuller, 1960, 83). We can argue that it is a conclusive level of strategy as a whole, or, following Mearsheimer (2010, p. 17), "the relationship between military means and international commitments." But the concept is hard to define since it is so intangible and only a few states have a properly thought-out let alone written grand strategy (Luttwak 1987, p. 178). According to Liddell Hart's (1932, p. 83) definition the role of grand strategy is "to co-ordinate and direct all the resources of a nation towards the attainment of the political object of the war—the goal defined by national policy." This is a simplified approach, since it focuses on action toward attaining political goals and the war effort of the nation. Liddell Hart emphasized the context of war and that is, indeed, the pinnacle of grand strategic activity, but most important grand strategical moves are conducted during peacetime. Not all interests of a state are worded in strategy papers, especially if the purpose is to exert power on other states. Grand strategy includes "the totality of what happens between states and other participant in international politics" (Luttwak 1987, p. 178). Armed forces are a strategic tool that can be employed for grand strategic purposes. As Bucholz (2001, p. 5) noted, the purpose of the armed forces is not to manage the day-to-day problems in the world. They exist to deal with future contingencies. The military works for political purposes and thus soldiers expect instructions. A commander

> is dedicated to a career that requires him to brood on the problems of war, in which activity he finds himself with very little civilian company. He does not have to be persuaded of his need for seasoned political guidance; the problem is rather one of making such guidance available to him on appropriate occasions and at appropriate levels. (Brodie 1959, p. 9)

A lack of understanding concerning the restrictions and potential of armed force makes it difficult to employ it optimally and in time. After WW II a war between two major powers as sole belligerents became a thing of the past since that numerous other nations would be drawn in immediately. Thus the "higher direction of war" becomes a problem of coordinating the efforts of allied nations (Martel 1945, p. 39). For the politico-strategical cooperation to work, the politicians require a thorough understanding of their military tool. Goltz (1906, p. 77) has argued how "we may unhesitatingly lay down the maxim that without a sound policy success in war is improbable."

Policy directs and guides warfare, and for the commander-in-chief, to have a clear idea what the outcome should be, a clear statement of the political end-state is required. The U.S. lost the Vietnam War even if it won every battle because of the lack of clear and immutable political will and statement of objective. Since the requirements for the armed forces were constantly changing, the military was unable to determine a desired end-state that would have tied battles into a campaign to fulfill the strategic objective. However, as Goltz noted, victory without sound policy is merely improbable and not impossible. When power-ratios of belligerents are imbalanced enough, even with a floundering political objective and resulting inability to plan strategically, the war may end in victory. But it is likely to become a drawn-out affair lasting much longer and consuming more resources than would have sufficed.

Old maxims of strategy still provide worthy guidance to the commanders and politicians of today. Leo VI (2010, p. 597) advised to "make a law for yourself to look first to the end of the course of the war and only then to begin it. It is despicable and mistaken to make a movement in war and then to have the army turn back again." If the end-state of the war has not been predetermined, the execution of operations becomes erratic. This demoralizes the troops, erodes popular support for the war, and operations don't proceed linearly to strategic objectives. It is easy to determine that the goal of military strategy and operational art is to subjugate the enemy and either destroy his armed forces or make them surrender. Nevertheless, grand strategy and policy have to aim higher. "It is necessary to distinguish between national (or grand) strategy and military strategy. The former deals with the whole of foreign policy in both peace and war and integrates military factors with political, economic, and diplomatic efforts" (Leonhard 1991, p. 6–7). A strategist has to ask the question posed by Marshal Foch: *De quoi s'agit il?* (What is it all about?) There is no other worthwhile goal of war than peace. A war is not fought just for victory, since it has to create the peace that follows it (Slessor 1954, p. 8). To wage war for peace sounds oxymoronic, but many great thinkers agree. St. Augustine (1909, p. 225) wrote, that

> war's aim is nothing but glorious peace: what is victory but a suppression of resistants, which being done, peace follows? So that peace is war's purpose, the scope of all military discipline, and the limit at which all just contentions level. All men seek peace by war, but none seek war by peace. For they that perturb the peace they live in, do it not for the hate of it, but to shew their power in alteration of it. They would not disannul it, but they would have it as they like; and though they break into seditions from the rest, yet must they hold a peaceful force with their fellows that are engaged with them, or else they shall never effect what they intended.

Peace is the ultimate desire of every warmonger, but a peace of their own choosing and one they have been designed or dictated. To postpone the next war, peace should be acceptable to all warring factions. Thus a stable and lasting peace can only result from a war that has ended in reasonable terms. The failure to comprehend this led to WW II soon after WW I. "WW I was the purposeless war, which no one seemed to know how to prevent and which, once begun, no one seemed to know how to stop" (Brodie, 1959, p. 55). Germany was devastated by the peace terms perhaps even worse than by the war itself. Yet, in less than two decades the world was aflame again. The harsh Treaty of Versailles failed to secure peace but laid foundations for a new war by sowing "*the dragon's teeth*" (Fuller 1948, p. 20). The failure of the treaty was

> to reduce a nation to a state of idiocy or of anarchy only means that it will be deprived of the power of fulfilling its contract—the terms laid down in the peace treaty. And if these terms are not fulfilled, then, from the point of view of policy, the war will, to a great extent, have been fought in vain; for policy should aim at attaining a more perfect peace than the one unhinged by the outbreak of hostilities. Conversely, the contract must be reasonable; for to compel a beaten foe to agree to terms which cannot be fulfilled is to sow the seeds of a war which one day will be declared in order to cancel the contract. Thus the national object is a better peace, and the means of attaining it is the conquest of the will of the hostile nation. (Fuller 1926, p. 71)

When WW II was nearing its inevitable conclusion, the demands of the victors were just as draconian and the prospect of severely punitive peace terms prolonged the war far beyond its termination-point had the Allies not demanded unconditional surrender (Kesselring 1989; Fuller 1948, p. 275). The humiliation of Germany was seen as just punishment for the aggressor, but it had disastrous consequences. An intolerable state of peace can be casus belli. War should be conducted with minimum sufficient amount of damage on the enemy to increase the odds of a lasting peace. Sherman held a completely contrary view to this. For him the American Civil War was a result of a criminal rebellion and thus he was an advocate of a violent and relentless war fought

> till all traces of the war are effaced; till those who appealed to it are sick and tired of it, and come to the emblem of our nation, and sue for peace. I would not coax them, or even meet them half-way, but make them so sick of war that generations would pass away before they would again appeal to it. (Sherman 1890, p. 368)

His vision became reality and the war was extremely harsh to both belligerents and destructive for the South. It was not an isolated example, since the dominant tendency of the nineteenth century and the first half of the twentieth was to fight wars to their bitterest ends (Stoker 2010, p. 6; Greenhalgh 2011, p. 479). This led Fuller to complain that the prevailing idea in the war was "to destroy each other, and so blinded were they by the means that they could not see that in the very act they were destroying themselves, not only during the war but in the peace which must someday follow the war" (Fuller 1923, p. 29). The same mistake was repeated in WW II (Fuller 1948, p. 27). The nature of war is destructive and chaotic, but policy has to restrain the use of force in warfare. Warfare cannot be allowed to escalate to a level when political goals become unreachable.

> A mad war can only lead to a mad peace, and to fight a war is such a way that an unprofitable peace is a certainty is clearly idiotic. It is this idiocy which, throughout history, has made the military mind so dangerous an instrument of government, for the warrior takes to destruction like a duck takes to water—chaos is his element. (Fuller 1946, p. 208)

The commanders-in-chief or the operational artists are not utterly innocent in creating these idiocies, but it is the task of the soldier to destroy and the task of the politician to rein him in. According to Napoleon, "the government must have entire confidence in its general; allow him great latitude and only provide him with the aim he should attain" (Cited in Fuller 1961, p. 45). Commanders are required to defeat their enemies as quickly and completely as possible and indirectly make the enemy willing to terminate the war. The commander-in-chief should have only one objective: the destruction of the enemy. The political leadership is responsible for setting, reevaluating, and reforming the political objectives of war continuously. Until politicians determine that suitable conditions for peace negotiations have been reached, the operational artists continue their own tasks to the best of their abilities (Seeckt 1930, p. 46). If the soldier strives for chaos and destruction he is only doing his duty. Yet it is not always the armed forces that yearn destruction, but the people. That the hopes for a better peace

> are rarely fulfilled is due to the ignorance and uncontrolled passion, and these fatal conditions are apt to be more marked on the side that is forced into the war in self-defence. It is the peaceful nations, above all, who need to learn that moderation in war is the best guarantee of subsequent peace. (Liddell Hart 1946, p. 76)

Once aroused from its peaceful slumber a democratic nation is a formidable enemy and the passions of the people fuel the destructive and punitive characteristics of the war. People can be very bellicose indeed (Howard 2008, p. 36–37). As Svechin (1992, p. 85) wrote, "mistaken policies will also bear the same pitiful fruit in war as they do in any other field." Policy can be supportive of or in juxtaposition to the military objectives but the strategist needs to follow the dicta of policy in his conduct of war since "political goals are not some kind of abstract digression for the strategist; they defined the main directions of the war" (Svechin 1992, p. 100). The politician should be coolheaded, remain above the situation, and keep his task in mind.

The oft-repeated claim that democracies would be more averse to war than other types of governance is unsupported by historical evidence. Based on the testimony of history, "the motive force of democracy is not love of others, it is the hate of all outside the tribe, faction, party or nation. The 'general will' predicates total war, and hate is the most puissant of recruiters" (Fuller 1961, p. 41). Today it is presupposed that liberal democracies have a certain structural attitude to war and their conduct of war rises from general parameters within which the liberal society works and thinks. This supposedly creates a "liberal democratic way" for conflict (Howard 2008; Gat 2011, p. 36–37). However, most Western societies have been spared from devastating war for decades and even liberal passions might flame up should there be a relapse (Dillon-Reid 2009, p. 111) "Pure violence, like fire, can be harnessed to a purpose; that does not mean that behind every holocaust is a shrewd intention successfully fulfilled" (Schelling 2008, p. 9).

Svechin (1992, p. 100) argued that the idea of giving "full power to a chosen military leader is an obsolete formula which never reflected any kind of reality." Exception proves the rule and we must remember that in their time Alexander and Napoleon were sovereign in strategy and politics. They were not given full power but took it and were free of political restrictions. Napoleon executed his strategical stratagems through his operational art. Based on this Jomini (1992, p. 49) wrote that "the prince should receive an education both political and military. He will more probably find men of administrative ability in his councils than good statesmen or soldiers; and hence he should be both of the latter himself." Napoleon was uniquely well poised to combine the characteristics of statesman and soldier. Egotism was the primary factor that made Napoleon one of the greatest captains of history, but simultaneously a reason why he couldn't build a lasting legacy (Moran 2007, p. 102). "In modern times the conduct of war has ceased to be a matter for the individual 'captain', combining the roles of commander and strategist; it has passed into the hands of government" (Tukhachevsky, cited in Simpkin 1987, p. 103). Especially during the Third Wave fusion of politics and strategy under the military would have dire consequences (Gray 2007, p. 51).

Politics must direct the war effort until the moment when planning phase is replaced by execution. Hence events accelerate and political control gradually recedes into the background. When the peak of operational activity in the war is reached, events follow each other too rapidly for detailed political control. Strachan (2011b, p. 507) has described this by arguing that "the dynamic and reciprocal nature of war shapes strategy more than strategy shapes war." As our civilizations have developed toward the Third Wave the accelerating speed of life has begun to shape the ways in which politics can supervise the military. Too close political supervision of military operations doesn't mean making wrong decisions but making them too late. The problem is that

> all independence, rapid decision, and audacious risk, without which no war can be conducted, ceases. An audacious decision can be arrived at by one man only. [...] If the highest military authority is not with the army, then it must allow the commander a free hand. War cannot be conducted from the green table. Frequent and rapid decisions can be shaped only on the spot according to the estimates of local conditions. (Moltke 1993, p. 77)

Setting and adjusting the limits for the use of force are political tasks, but the government should not dictate how the armed forces fight, only define the means at their disposal. Arguing that once the army has been set in motion it should be left to deliver the desired policy objectives is un-Clausewitzian. Rather there is a more pressing need for politicians and soldiers to re-engage so that war and policy are brought into step (Strachan 2013, p. 55). Today military strategy has become "*the diplomacy of violence*" (Schelling 2008, p. 34). In the past the situation was less nuanced. Moltke considered warfare as a part of *Realpolitik* and thus policy cannot be separated from strategy, for politics uses war to attain its objectives and has a decisive influence on war's beginning and end. Politics does this in such a manner that it reserves to itself the right to increase its demands during the course of war or to satisfy itself with minor successes. [...] Strategy thus works best in the hands of politics and only for the latter's purposes. But, in its actions, strategy is independent of policy as much as possible. Policy must not be allowed to interfere with operations. (Moltke 1993, p. 36)

Moltke wanted to keep the armed forces factually separated from political chain of command (Bond 1998, p. 75–76). The military apparatus of the state could thus perform its appointed duty without political guidance or restrictions to its operations. "In no instance must the military commander allow himself to be swayed in his operations by policy considerations only. He should rather keep military success in view. What policy can do with his victories or defeats is not his business" (Moltke 1993, p. 36). If we view war as a continuation of politics with additional means, we must understand that the relationship will be altered during wartime since "the center of gravity of the political struggle is transferred from the non-military to the military. Politics exchanges the pen for the sword and new relations and laws become operative" (Sokolovsky 1963, p. 22). As operational art becomes for a moment the grammar of violence as part of political dialogue, the ability of the operational artist to manipulate time has political consequences as well.

In the course of time, the ability to understand politics in addition to strategy and operational art has become a prerequisite of contemporary operational artists (Clark 2001; Franks 2004; McChrystal 2013). "The modern general must think seriously about many more factors that fire and maneuver. Political, economic, and cultural elements exist not only as constraints, but as positive opportunities to gain the advantage in conflict" (Leonhard 1998, p. 58).

1.6 Operationalizing the Art of War

> War differs from battle. War is defined as going from the beginnings to the end, that is, the cessation of hostilities, and it includes many battles in its course. Battle is defined as a partial war that occurs frequently in the course of the entire war, and its cessation does not always bring about the end of the war but, as need requires, battles can take place two or three times or more often in the course of the entire war. (Leo VI 2010, p. 385)

As Leo's *Taktika* shows, the entire idea of warfare was a lot simpler in the past. War itself, the period from beginning of hostilities to their cessation, still fits the definitions, but it has become immeasurably more complicated. A battle is no longer even "partial war" but only a building block in one of potentially numerous operations that join together into the multi-leveled concept of war. The taxonomy of the levels of war today is almost universally accepted. Warden (2000, p. 1) listed the levels as grand strategic, strategic, operational, and tactical. Often the highest level is discussed with the more common term 'politics.' Strategy is divisible to operational art and branch-specific subtypes of operations and further into different tactics. The principles that guide war as a phenomenon can be applied to its constituent units. Mahan (1999, p. 65) argued that the principles vary on all levels of the art of war depending mostly on the change and evolution of weapons and technology, but that the changes are smaller in strategy than they are on the level of tactics.

Thus, identifying principles and estimating their validity is easier on the higher levels. Fuller ties the different levels of warfare together by arguing that the principles of war "are as applicable to strategy (operations in plan) as to tactics (operations in action), two terms which should never be separated by a bulkhead, because their components flow into each other and together constitute the art of war" (Fuller 1960, p. 293). The art of war has different components of specific tactics and forms of operational art that influence each other. As an example the Western operational art today is commonly agreed to consist of "joint" operations in which each branch of service supports and complements each other for a unified purpose. Today planning and execution of large-scale operations that combine air, land, and sea forces are the essence of operational art (Matheny 2012, p. xiii).

Time and its complex relationships, dependencies, and influences to warfare are elusive to analyze. Grand strategic viewpoint may span all actions of a nation-state for decades ahead; strategy generally concerns itself with years and operations with months. For Gray (2007, p. 70) time is the least forgiving dimension of strategy but seldom discussed in depth. Most authors quickly note its importance and move on to tractable matters. In strategy once time is lost, it cannot be recovered. But how is time manifested in operational art?

Miyamoto Musashi saw strategy as the professional skill of the soldier regardless of artificially constructed levels. He saw "the art of the advantage" to be the abstract core of the art of war and it refers to the science of strategy in general (Cleary 2005, p. 24–25). At thirty he began to study strategy or the Way of War and at fifty he finally began to understand it. The Way has to be known to every soldier, but it remains the primary concern of the leader. The Way is two-fold; that of the sword and that of the pen, and the warrior needs to follow both paths (Musashi 1995, p. 4447). There is a time and place for use and non-use of weapons (Musashi 1995, p. 56). The former is the realm of operational art. One can become a master of strategy by practicing to understand the plots of the enemy as well as his strength and means (Musashi 1995, p. 77). The Way begins with knowing the basics thoroughly and on this foundation a strategy is built. Thus, to be able to conquer one means the same as vanquishing ten million. To know one thing properly equals knowing ten thousand, since the strategist includes the small into the vast (Musashi 1995, p. 53). Musashi shared the idea of the nature of war with Clausewitz who later formulated in more detail:

> War is nothing but a duel on a larger scale. Countless duels go to make up war, but a picture of it as a whole can be formed by imagining a pair of wrestlers. Each tries through physical force to compel the other to do his will; his immediate aim is to throw his opponent in order to make him incapable of further resistance. War is thus an act of force to compel our enemy to do our will. (Clausewitz 1989, p. 75)

A general wields an entire army and a swordsman only his *katana*. The battle is still a duel (Smith 2008, p. 16), and as Jomini pointed out, the difference between levels of war is mostly the viewpoint and the direction of the planning process. "In strategy the general begins with the planning of the entire war and campaigns within it and descends to the details. A tactician begins at the grass-roots level and ascends to combining details together to create a system of forming and handling an army" (Jomini

1992, p. 66). In Musashi's writings all actions of a warrior are a part of strategy. His idea was more inclusive than the division into levels today. To conceive of war as a true art and to learn to appreciate its terrible beauty in its multiple forms, the thinking of Musashi is beneficial for the strategist, operational artist, and tactician alike (Cleary 2005, p. 79–80).

Many of the principles Musashi used to describe a swordfight can be applied to wider context of battle or campaign. When the warrior picks up his sword, he should not think about striking or slashing, but only of continuing his movement to the penultimate strike (Musashi 1995, p. 67). Using the swordsman as an example Musashi illustrates strategy for an army or a state. All actions should be conducted to together result in final victory. The same applies to his concept of "striking the enemy with one timing." When close to the enemy one needs to strike without moving one's position, without pausing to gain courage, striking rapidly when the enemy is still undecided. It means striking at the blink of an eye, before the enemy knows whether to back off or perform his own strike (Musashi 1995, p. 67). Time is too important to be given to the enemy.

The early indust-real levels of warfare consisted of only strategy and tactics and the two differed in their treatment of time. "There are essential differences in the goals and in the manner of command of large and small units. What is necessary for the former is not right for the latter. Space and time have for each a different meaning" (Moltke 1993, p. 130). Space and time are crucial elements of both. The details of terrain, for example, become more meaningful the lower down the ladder we descend. One would plan his entire offensive differently in Sahara than in the glaciers of the Himalayas. The same applies to time. In tactics the units of time are smaller than in operational art, but in both consuming as little time as possible was important in indust-real warfare.

In the ancient world the relationship was simpler. Strategy was the art of the general and tactics guided the actions on the battlefield (Bülow 2013). Both the Greeks and the Romans excelled in these two. But the concepts have not remained unchanged over time. During the Dark Middle Ages both tactics and strategy degenerated too badly to be considered as art (Delbrück 1990c, p. 323–329). Yet they were revived with the Enlightenment and refined during the indust-reality. Prior to WW I the Soviet military art spoke only of two main elements. They were "strategy as teaching on war, and tactics as teaching on battle" (Isserson 2013, p. 13). Lately new terms have sprung up and expressions such as 'operational-strategic' or 'theater-strategic' are commonly used, but we stick to the basics. Jomini provided a taxonomy of the different branches of the art of war. He divided them into five purely military ones; Strategy, Grand Tactics, Logistic, Engineering, and Tactics. The sixth branch he named "Diplomacy in its relation to War" and argued that while it is more the statesman's business than the soldier's, it needs to be understood by commanders since "it enters into all the combinations which may lead to a war, and has a connection with the various operations to be undertaken in this war" (Jomini 1992, p. 13). Most of Jomini's terms are self-explanatory, but he used his "Grand Tactics" as we would use "operational" today.

The name is fitting, since operational art is to some degree tactics on a higher level. According to his definition "Grand tactics is the art of making good combinations preliminary to battles, as well as during their progress" (Jomini 2007, p. 131). Temporally the realm of grand tactics begins from planning how strategy could be turned into action and extends through the period of actual operations until the strategic goals are reached. It doesn't concern battles that belong to the realm of tactics. Prior to battle grand tactics attempts to create force concentrations that enable victory and after the battle creates its "combinations" based on the outcome. To carry out the dictates of strategy and grand tactics troops needed to engage in battle.

> The maneuvering of an army upon the battle-field, and the different formations of troops for attack, constitute Grand Tactics. Logistics is the art of moving armies. It comprises the order and details of marches and camps, and of quartering and supplying troops; in a word, it is the execution of strategical and tactical enterprises. (Jomini 2007, p. 51)

Strategy thus outlined how the entire military might of a nation was to be used. Grand tactics consisted of positioning and maneuvering an army on the battlefield and logistics was for Jomini the art of getting the armies to the battlefield and supplying them. In other words, "Logistics comprises the means and arrangements which work out the plans of strategy and tactics. Strategy decides where to act; logistics brings the troops to this point; grand tactics decides the manner of execution and the employment of the troops" (Jomini 1992, p. 69). Once troops have reached the battlefield, tactics govern them for the duration of the fighting. Meanwhile grand tactics plans their post-battle re-employment.

The first to use the expression 'grand tactics' was Jacques Antoine Hipolyte de Guibert (1773, p. 5). From his perspective it consisted of assembling forces, unifying them and executing grand maneuvers with them. Thus grand tactics was an art of executing, combining, and leading these grand maneuvers. It was a compilation of all military knowledge and Napoleon, for example, shied from using the word 'strategy' but preferred 'grand tactics' for what he called the "higher parts of war" (Vego 2009, p. I–5; Strachan 2011a, p. 98; Strachan 2013, p. 29).

'Operational art' is a relatively new concept in the anglophone art of war. As late as 1965 in France it was argued that alongside tactics and strategy the only other level of war is logistics as "the science of supply and movement" (Beaufre 1965, p. 22). 'Operations' were "*the sum total of the dispositions and manoeuvres*" required to get armies to oppose each other on the battlefield (Beaufre 1965, p. 59). Operational art can be conceived of as a modernized version of grand tactics. If the function of strategy was for Gray (2010, p. 15) to hold a bridge between politics and military action operational art can be conceived of as another bridge that connects the tactical to the strategic level. There is overlap at each end. The boundaries of operational art are not clearly drawn, but its primary meaning is to act as the level of thought and activity that guides the conduct of operations and connects strategy to tactics enabling us to treat art of war as a unified whole. In agrarian times and early indust-reality there was no need for such a concept. The great captains of the past forged their strategy and policy through battles. An entire war could be composed of one single clash of arms that would decide the destiny of the loser. Tactical level continued to have a direct link to strategy as late as the Napoleonic Wars.

After Napoleon, there came a change in the structures of the armies and governments. Indust-reality brought along the cumbersome state machinery as an organizer and the concept of bureaucracy entered every function of the society. Even if armies of the time had somehow managed to avoid adopting processes of the society, the internal complexity of mass armies created a gap between strategy and tactics and a need for a new level to fill the gray zone between the two. At least from the nineteenth century onward art of war has had different manifestations on tactical, operational, and strategic levels (Mattheny 2012, p. 15). Creation of a separate level of operational art to govern campaigns or operations was a byproduct of the indust-real concept of nation in arms. The development was simultaneous and occurred as a reaction to altered conditions. "The term the French adopted and the British followed was *la grande tactique* (grand tactics). On the German side the word chosen was *operativ*, quickly borrowed by the Russians as *operativnyi* and now rendered in English as 'operational'" (Simpkin 1985, p. 14). Fuller, for example, never wrote about operations or operational art with those specific expressions. He wrote about grand tactics, but taking the perspective of operational art helps to understand his theories (Reid 1998, p. x–xi, 17–18).

For the Soviet theorists, the span of operational level covered military hierarchy from army group or army to corps and occasionally division (Glantz 1991, p. 10). The term 'operational art' was first coined by Aleksandr Svechin in 1922 (Vego 2009, p. I–23). He argued that it was a critical conceptual linkage between tactics and strategy since combat actions are not self-contained but basic material out of which an operation is created. To gain victory in war one had to follow a path "broken down into series of operations separated by more or less lengthy pauses which take place in different areas in a theater and differ significantly from one another due to the differences between the immediate objectives one's forces temporarily strive for" (Svechin, cited in Vego 2009, p. I–24). In order to gain strategic objectives first individual battles had to be joined into operations and they in turn combined into a unified whole. Isserson (2013, p. 10) noted that "instruction about contemporary operations is insufficiently worked out and remains the least elaborated aspect of military art." The Soviet thinkers took a head-start in conceptual development of operational art because they first perceived that changes in the composition of armies demanded a re-evaluation of the tenets of the art of war. In Russia, the theory of operational art was born because it was understood that "the ultimate war aims cannot be accomplished by one blow, our military theory considered it necessary to conduct a series of integrated campaigns and operations" (Sokolovsky 1963, p. 129).

While operational art was 'invented' in the Occident, the concept is universal. In translations the 'operational' level is often referred to by talking about 'campaigning.' Mao and Giap expressed similar understanding of an intervening level between strategy and tactics in the context of war. Mao argued that "*the science of strategy is to study those laws for directing a war that govern a war situation as a whole. The task of the science of campaigns and the science of tactics is to study those laws for directing a war that govern a partial situation*" (Mao 1963, p. 7980). Strategy explores war as phenomenon while science of campaigns concerns itself with its components. In Vietnam "military art determines the organic relationship and interaction among strategy, campaign, and tactics, which are the components that make up this art, and it correctly determines the role of each component" (Giap 1970, p. 75). Strategy creates conditions that allow objectives of both campaign and battle to be reached. Fulfilling tactical tasks on the battlefield solves operational problems related to the campaign and creates a basis to achieve strategical objectives (ibid.). Operational art or the art of campaigns completes the three-tiered structure, but "tactics are inseparable from strategy" (Giap 1968, p. 40).

Intellectually operational art is considered to govern "the planning level of war that constructs campaigns and major operations in order to accomplish the theater goals articulated at the strategic planning level; the intermediate planning level that integrates tactical efforts and events into a campaign" (Leonhard 1991, p. 8–9). Operational art is thus both the practical execution of strategic plans and planning the creation of sequential operations and campaigns. Strategy sets the operational level end-states as intermediate goals and shortest logical links on the way to the strategic goal itself (Svechin 1992, p. 317). To put the task of operational art in as simple form as possible, "What the strategic plan demands, operational art must supply. Likewise tactical objectives must slavishly submit to the operational plan" (Leonhard 1991, p. 10).

Clausewitz has defined strategy as the use of battle for the purposes of war. In his time strategy was the art of the commander (Strachan 2013, p. 11). It was "the combination of individual engagements to attain the goal of the campaign or war" (Clausewitz 1989, p. 349). This is close to what we today call operational art. According to Matheny (2012, p. 9–10) both Clausewitz and Jomini wrote about operational art even if they called it strategy. This is overstating the case. It is, rather, that in time strategy moved beyond actual battles. Strategy today creates links between war and policy (Strachan 2013, p. 57), and operational art became the medium to re-connect tactics and strategy. Strategy has expanded into the realm of politics and distanced itself from tactics, leaving the intermediate area to be called 'operational.' Clausewitz wrote in the context of Napoleonic field armies and the need to combine the effort of different branches of service was different from today. Art of war was simpler and this led Clausewitz to highlight the role of tactics over the lowest levels of what he called strategy. For Isserson (2013, p. 35) tactics and operational art "differ only in scope and dynamics. They not only co-exist during combat actions, but they organically flow into one another." Tactical successes should become operational ones, since tactical, operational, and strategic levels exist in the same time frame and blend. Battles are the building blocks.

> If we know how to fight and how to win, little more knowledge is needed. For it is easy to combine fortunate results. It is merely a matter of experienced judgment and does not depend on special knowledge, as does the direction of battle. (Clausewitz 1989, p. 349)

Clausewitz noted that it is easier to initiate the next battle after a victorious one and operationally build upon success when compared to the problems one encounters in rallying troops after a lost battle. Furthermore, tactical command in battle requires specialized skills and detailed knowledge of the troops. In operational art or strategy much of the detail is omitted from the bigger picture of war. But there can be no effective operational level victory without excellence on the battlefield. "Only great tactical successes can lead to great strategic ones; or as we have already said more specifically, tactical successes are of paramount importance in war" (Clausewitz 1989, p. 228–229). Strategy may be successful, but

> in strategy there is no such thing as victory. Part of strategic success lies in timely preparation for a tactical victory; the greater the strategic success, the greater the likelihood of a victorious engagement. The rest of strategic success lies in the exploitation of a victory won. (Clausewitz 1989, p. 363)

This ability to exploit success is no meager task, since battles expend energy and resources and in the course of the campaign "the strain will gradually increase and in the end resources may be exhausted. Time is thus enough to bring about a change unaided" (Clausewitz 1989, p. 598).

Therefore, in order to avoid stagnation one must not operationally allow for a moment of relaxation but push onward after each tactical victory, because then the enemy is most vulnerable. Every battle has the potential to alter the future course of the campaign. In this as many other things Moltke followed Clausewitz (Howard 1983, p. 59) by writing that

> the tactical result of an engagement forms the base for new strategic decisions because victory or defeat in battle changes the situation to such a degree that no human acumen is able to see beyond the first battle. (Moltke 1993, p. 92)

The outcome of a battle dictates what future actions remain viable. True operational art lies in flexibility and the skill to recover after a loss. To build a victory on the foundation of a lost battle and return on the operational line leading to the achievement of set goals is the pinnacle of operational art. Knowing only how to fight and win doesn't guarantee success, since much in warfare depends on circumstances and even luck. Knowing how to make the best of what is at hand at any given moment and utilize resources optimally to reach objectives is the hard part of operational art. Operational art is the art of campaigning. Ultimately victory in war is not won by winning all the battles, as the U.S. proved in Vietnam. Every battle has to be a part of a chain of operations and operations combine together to achieve the strategic goal. After the humiliation in Vietnam, American acceptance of the concept of operational art started to increase and a conceptual leap from winning battles to winning operations occurred (Adamsky 2010, p. 60). Operational art is popular today but as late as the 1990s amidst the hype around it Leonhard (1991, p. 5–6) wrote that "the American army is generally untrained in operational art, despite our promiscuous use of the term of late."

Operational art is a peculiar art form, because it is not for everyone. Marshal Foch wrote that war is not "an art for dilettante, a sport. You do not make war without reason, without an object, as you would give yourself up to music, painting, hunting, lawn tennis, where there is no great harm done whether you stop altogether or go on, whether you do little or much" (Foch 1920, p. 13). Operational art is too serious in its consequences to be dabbled with. A symphony, a painting, or a sculpture can be left unfinished when the fancy no longer grips the artist. War is an artwork that has to be finished with blood, sweat, and tears. One can plan his actions and meditate on them, but at some point the mental Rubicon either has to be crossed or one must stand down. "Every attack, once undertaken, must be fought to a finish; every defence, once begun, must he carried on with the utmost energy" (Foch 1920, p. 340).

1.7 The 'Eternal' Principles of War

Warfare is supposedly guided by the so-called 'principles of war' and we will focus on the ones that are concerned with time and temporality. Every officer is taught these principles in one form or another and the old maxims and principles are constantly discussed. Gray (2007, p. 60) argues that Sun-Tzu, Clausewitz, and Thucydides are endowed with timeless strategic ideas that can be improved upon but not updated. Thucydides and Sun-Tzu devoted their attention to issues that precede war and for them diplomacy was the way to win wars before they started and attain bloodless victories. For Sun-Tzu (1993, p. 157) war was "the greatest affair of the state, the basis of life and death, the Way [Tao] to survival or extinction. It must be thoroughly pondered and analyzed." Clausewitz's work instead focused on the actual art of waging war (Handel 2001, p. 33). Even if conditions of war vary from age to age, as Mahan (1999, p. 57) wrote, "there are certain teachings in the school of history which remain constant, and being, therefore, of universal application, can be elevated to the rank of general principles." Principles occasionally have to be deduced from the texts. Clausewitz is an example of a strategist whose text contains many references to the need for a system of principles but who "never delivers them in a way designed to be learnt by the parrots of military crammers and spoon-fed examinees" (Strachan 2013, p. 203). Even when principles are listed, their meanings might remain elusive. For Fuller (1923, p. 40)

> the principles in themselves are not worth the paper they are written on, for they are but mere words strung together in a certain order. Their value lies in their application, and this application depends on the thousand and one conditions which surround the elements of war during operations. What are these conditions, for without knowing them it is manifestly impossible to apply the principles? Conditions are innumerable and ever changing, but the following are some of the most important: Time, space, ground, weather, numbers, training, communications, supply, armament, formations, obstacles and observation.

Some of the conditions Fuller listed keep their meaning intact while others are specific to time and culture. There are commonalities as well as distinctive differences between cultures (Scobell 2011). Principles remain empty slogans and catchwords unless they are adapted to the particular temporal and cultural context. Dragomirov claimed that the principles "are within the reach of the most common-place intelligence; but this does not mean to say that such an intelligence would be able to practice them." (Cited in Foch 1920, p. 115). They are easy to memorize, but their employment requires a highly specialized form of intelligence in a commander. Failure to comprehend the fluid and changing context in which the principles are applied has led some theorists to question them. Leonhard (1998, p. 8) argued that the principles "have in fact changed many times even in the brief history of our country. Other nations—some close allies—disagree with our list of principles, some substituting their own lists, others claiming that there are no valid principles."

The idea of 'eternal' principles is quite recent. In natural sciences systematical models were built to reveal universal principles that dominate phenomena (Handel 2001, p. 53; Gat 2001, p. 11). A philosophic contemplation of war was countered with tenets from the harder sciences.

Physics as an approach to art of war may have suited the mechanized age but as we still lack a 'quantum theory' of war it is easy to join Admiral Wylie (2014, p. 20) in his claim that the principles are "an attempt to rationalize and categorize common sense." War is an art and not a trade or science. The principles need to be learned to be applied, but they are unlike the inflexible laws of Newtonian physics (Colby 1943, p. 6). They describe how war should be conducted for optimal performance and a greater chance of victory. They don't dictate the course of warfare. There are no laws of war that enable predicting outcomes with certainty since war is riddled with unpredictability and possible subsequent turns of fortune (Coker 2010, p. 171). Following the principles of war increases the likelihood of the commander to subjugate his enemy, but unthinking obedience to them doesn't guarantee success.

Every now and then old ideas and adaptations of the art of war are tested and some are proven untrue while others persist. As an example suffices the advice to avoid waging war simultaneously on two separate fronts. It is not a principle in its own right but a result of adhering to the principles of concentration of force and economic use of force. In the words of Jomini (2007, p. 25), "the celebrated maxim of the Romans, not to undertake two great wars at the same time, is so well known and so well appreciated as to spare the necessity of demonstrating its wisdom." Nevertheless, both World Wars saw Germany violate this principle and U.S. military strategy for decades focused on building a sufficient capability to fight wars on two separate theaters. Only in the aftermath of the Cold War and the War on Terror this idea seems to be on the wane (Kagan 2006, p. 152–159; Clark 2001, p. 47–49). Peter Paret (2009, p. 5) stated that "history is better at revealing than proving, and states do not interact in controlled laboratory conditions that allow comparisons of precisely equal elements. We must study war in the shifting reality in which it occurs." Thus the principles or any other dicta don't create a strict framework for warfare. For Fuller (1960, p. 293) principles are guidelines to be discarded if necessary. They are

> no more than pegs on which to hang our tactical thoughts. There is nothing irrevocable about them; sometimes they may be discarded with impunity; but as a study of military history will show, they should only be discarded after deep consideration. They are very important guides rather than principles.

But what is the usefulness of such principles of war? As Bernard Brodie (1959, p. 21) argued, they are derived from "the work of a handful of theorists, most of them long since dead. Their specific contributions to living doctrine are not widely known, because their works are seldom read. The richness of their ideas is but poorly reflected in the axioms which have stemmed from those ideas." All too often the original works are either not read or only partially assimilated and thus random maxims are borrowed and circulated. Clausewitz is perhaps the best example with the omnipresent misquote about war being a continuation of politics by other means. He never defined his "*Politik*" in any detail in those contexts when anglophones need to choose between 'politics' and 'policy,' and this causes misinterpretations (Bassford 2007, p. 83). Clausewitz created a broad analytic framework by taking war out of its military isolation and embedding it in society and politics (Paret 2009, p. 117) and this broadening of perspective has led to multiple misunderstandings throughout the years. Lenin understood the connections between war and politics similarly and built the communist principles of war around this relationship (Simpkin 1987, p. 24; Heuser 2002, p. 46–47). He stressed that "war is part of a whole, and this whole is politics. He also stated that war is a continuation of politics and that politics also 'continue' during war" (Sokolovsky 1963, p. 167). But the principles of war themselves are not political, because they relate to warfare and not the policy governing it. War is an extension of policy. It is politics *with* other, additional, means and the principles illustrate the use of these means. The principles of war are

> as applicable to sea warfare and to air warfare as to land warfare, irrespective of the differences in the three spheres of action in which these three modes of warfare take place, the spheres of sea, air and land. The ultimate objective is the same, namely, the maintenance of policy. The two great means are the same—offensive and defensive action, whether material, physical, or moral.

> The methods of potentiating these means are identical—concentration and economy, movement and surprise, and the ultimate co-ordination is the same—co-operation within fleets, armies and air forces and cooperation between them as parts of one single defence force. It is this co-operation which, I consider, forms the foundation of grand tactics, not as heretofore interpreted—the major battle plan of an army, or of a navy, or of an air force, but of an army, a navy and an air force intimately co-operating in order to attain a common objective—the maintenance of policy. (Fuller 1923, p. 216)

Many classics of the art of war were written about land warfare, and the acceptance of war being a joint employment of all services and unified expression of national policy emerged late. There are notable exceptions like Julian Corbett (1999, p. 167) who wrote that one should not study naval or land strategy as isolated concepts since "embracing them both is a larger strategy which regards the fleet and army as one weapon, which coordinates their action, and indicates the lines on which each must move to realise the full power of both." During indust-reality war was still mostly broken down to service-specific components. Often for the soldier war was

> a nest of pigeon-holes: strategy, tactics, organization, administration, etc., etc., each nest being crammed with pill-boxes—infantry tactics, cavalry tactics, artillery tactics, etc., etc. The danger underlying these uncorrected values is to be sought in the temptation to invest them with individual, that is separate, existences, and then, when combined action is demanded, to produce a mixture of values in place of a compound. (Fuller 1923, p. 212)

Warfare should not be conceived of as a mix of different tactics or strategies, but a phenomenon that first needs to be understood as a whole and only then dissected into more palatable pieces. War is a compound where different elements blend to create something from which the ingredients cannot be mechanically separated. Fuller (1923, p. 216) wrote that,

> whether man fights on the land, or on the sea, or in the air, the elements of war are the same, namely: moral, weapons, movement and protection; consequently, whatever mode of war is to be examined, in these elements we find a common denominator to all three forces. If this be accepted as correct, then I see no reason why warfare as a whole should not be treated as one subject.

Tactics, operations, and strategy are always tied to their temporal, political, and cultural contexts. They assume a certain form because of the time when they are applied and by whom. Thus one can follow Clausewitz and accept the singular and individual character of past wars to take into account the differences and generalities between them in a loose framework of historicism. Wars differ because their contexts are different (Paret 2009, p. 121). It is not enough to speak of temporally different contexts but even cultural or national ones. As von der Goltz (1906, p. 77) put it, "a writer upon strategy and tactics ought to treat his subjects as national strategy and tactics; for only such teaching can be of real service to his country." What applies to one country may be inapplicable to another. Thus, the resources at the disposal of a superpower are different than those that dictate the tactics, operations, and strategy of a small state. Operational art turns warfare into an art instead of being merely planning on the highest and fighting at the lowest levels. It is "primarily concerned with how to achieve the strategic end of the war with the forces allotted" (Warden 2000, p. 2). Operational art discusses those elements of the art of war that are altered to best suit the circumstances of their application. Mao (1963, p. 77–78) agreed, arguing that

> the different laws for directing different wars are determined by the different circumstances of those wars—differences in their time, place, and nature. As regards the time factor, both war and its laws develop; each historical stage has its special characteristics, and hence the laws of war in each historical stage have their special characteristics and cannot be mechanically applied in another stage. (...) In studying the laws for directing wars that occur at different historical stages, that differ in nature and that are waged in different places and by different nations, we must fix our attention on the characteristics and development of each, and must oppose a mechanical approach to the problem of war.

Thus operational art is the art of thinking how to best apply the principles of war in a particular circumstance. An understanding of these principles is a prerequisite of creating a national way of waging war. Operational art is universal, but operations are conceived differently in each specific case. Unless operational art is a subject of constant study it withers and dies. If its teachings don't influence national tactics they become standardized. "Thus there remain nothing but customs, the principles of which are unknown to us" (de Saxe, 1944, p. 18). Customs are the primary opposition of progress and evolution of the art of war. Likewise, as Liddell Hart (1932, p. 77) wrote, "standardization is the curse of modern armies and modern thought." The deeper first mechanization and then high-tech weaponry of the information age have penetrated into the art of war, the more freedom of thought is restricted when the tools of war start to dictate how it should be waged. In the agrarian age the relationship of the man and his tools was different from the indust-reality and in our information societies the relationship needs to be rethought again. The tools and their users have in each age a different impact on the relative importance of time and its meaning in operational art. Treatment of armed forces as a machine striving for uttermost economy of time and other resources was a byproduct of indust-reality, but efficient use of all assets remains important.

> Wastefulness is an ugly part of man, and the practice of warfare is particularly vulnerable to it. The dynamics that dominate warfare—uncertainty, fear, error, miscalculation, and often incompetence—lead to uneconomical practices in war. At times, inefficiency leads only to time lost, treasure wasted, or equipment poorly used. But all too often, human blood is the price of ineptitude. Of all endeavors of humankind, warfare has the potential to be the most uneconomical. (Leonhard 1998, p. 124–125)

In this chapter we have seen that only war itself is eternal and immutable; its character or outlook changes not only over time but from one context of applying the art of war to another. The groundwork for treating time and temporality as important building blocks of any masterpiece of the art of war has been laid, On each level of warfare, from politics to tactics, the perception and meaning of time and temporality may be different. The next chapter will burrow deeper into how exactly time can be utilized in operational art. In practical application of the art of war, time never exists in a vacuum, but becomes important only in conjunction with the other two essentials of operational art: space and force. Operational art in effect consists of creating a favorable equilibrium of this trinity to best suit each given occasion.

References

Adamsky, D. (2010). *The Culture of Military Innovation – The Impact of Cultural Factors On the Revolution in Military Affairs in Russia, the US, and Israel.* Stanford: Stanford Security Press.

Alberts, D. S., Garstka, J. J., & Stein, F. P. (2000). *Network Centric Warfare – Developing and Leveraging Information Superiority* (2nd ed. revised), CCRP Publication Series.

Bassford, C. (2007). The Primacy of Policy and the 'Trinity' in Clausewitz' Mature Thought. In H. Strachan & A. Herberg-Rothe (Eds.), *Clausewitz in the Twenty-First Century* (pp. 74–90). Oxford: Oxford University Press.

Bassford, C., Moran, D., & Pedlow, G. W. (2010). On Waterloo – Clausewitz, Wellington, and the Campaign of 1815. Clausewitz.com.

Beaufre, A. (1965). *An Introduction to Strategy – With Particular Reference to Problems of Defense, Politics, Economics, and Diplomacy in the Nuclear age.* New York/Washington: Frederick A. Praeger.

Bellamy, C. (2007). *Absolute War – Soviet Russia in the Second World War.* New York: Alfred A. Knopf.

Bond, B. (1998). *The Pursuit of Victory: From Napoleon to Saddam Hussein.* Oxford: Oxford University Press.

Brodie, B. (1959). *Strategy in the Missile Age.* Princeton: Princeton University Press.

Bucholz, A. (2001). *Moltke and the German Wars, 1864–1871.* New York: Palgrave Macmillan.

Caesar, G. J. (2010). *De Bello Gallico and Other Commentaries.* Miami: HardPress Publishing.

Clark, W. K. (2001). *Waging Modern War.* New York: Public Affairs.

Cleary, T. (2005). *The Japanese Art of War - Understanding the Culture of Strategy.* London: Shambhala Classics.

Coker, C. (2004). *The Future of War: The Re-enchantment of War in the Twenty-First Century.* Hoboken: Wiley-Blackwell.

Coker, C. (2010). *Barbarous Philosophers: Reflections on the Nature of War from Heraclitus to Heisenberg.* New York: Columbia University Press.

Colby, E. (1943). *Masters of Mobile warfare.* Princeton: Princeton University Press.

Coram, R. (2002). *Boyd, the Fighter Pilot Who Changed the Art of War.* New York: Back Bay Books.

Corbett, J. (1999). Some Principles of Maritime Strategy. In *Roots of Strategy* (pp. 149–262, Book 4). Mechanicsburg: Stackpole Books.

Davis, P. K. (2001). *Effects-Based Operations: A Grand Challenge for the Analytical Community.* RAND NDRI and Project AIR FORCE Publication. Santa Monica: RAND.

de Bloch, J. (1914). *Future of War in Its Technical Economic and Political Relations.* Boston: The World Peace Foundation.

De Gaulle, C. (1976). *The Army of the Future*. Westport: Greenwood Press.
de Saxe, M. (1944). *Reveries upon the Art of War*. Harrisburg: Military Service Publishing Company.
Delbrück, H. (1990a). *Warfare in Antiquity - History of the Art of War* (Vol. I). Lincoln/London: University of Nebraska Press.
Delbrück, H. (1990b). *The Barbarian Invasions - History of the Art of War* (Vol. II). Lincoln/London: University of Nebraska Press.
Delbrück, H. (1990c). *Medieval Warfare - History of the Art of War* (Vol. III). Lincoln/London: University of Nebraska Press.
Delbrück, H. (1990d). *The Dawn of Modern Warfare - History of the Art of War* (Vol. IV). Lincoln/London: University of Nebraska Press.
Delbrück, H. (1997). *Delbrück's Modern Military History*. (A. Bucholz, Ed. and Trans.). Lincoln/London: University of Nebraska Press.
Der Derian, J. (2001). *Virtuous War – Mapping the Military-Industrial-Media-Entertainment Network*. Boulder: Westview Press.
Dillon, M., & Reid, J. (2009). *The Liberal Way of War. Killing to Make Life Live*. London: Routledge.
Douhet, G. (1999), The Command of the Air. In *Roots of Strategy* (pp. 263–408, Book 4). Mechanicsburg: Stackpole books.
du Picq, A. (1987). Battle studies – Ancient and modern Battle. In *Roots of Strategy* (pp. 65–300, Book 2). Harrisburg: Stackpole Books.
Dupuy, T. N. (1987). *Understanding War – History and Theory of Combat*. New York: Paragon House Publishers.
Echevarria, A. J., II. (2000). *After Clausewitz – German Military Thinkers Before the Great War*. Lawrence: University Press of Kansas.
Echevarria, A. J. II. (2011). American Operational Art, 1917–2008. In J. A. Olsen & M. van Creveld (Eds.), *The Evolution of Operational Art – From Napoleon to the Present* (pp. 137–165). Oxford: Oxford University Press.
Foch, F. (1920). *The Principles of War*. New York: Henry Holt and Company.
Foertsch, H. (1939). *Nykyinen ja tuleva sotataito*. (Org. Kriegskunste Heute und Morgen). Porvoo: Werner Söderström Oy.
Franks, T. (2004). *American Soldier*. New York: Regan Books.
Freedman, L. (2006). *The Transformation of Strategic Affairs* (Adelphi Paper 379). London: The International Institute for Strategic Studies.
Fuller, J. F. C. (1923). *The Reformation of War*. London: Hutchinson.
Fuller, J. F. C. (1926). *The Foundations of the Science of War*. London: Hutchinson.
Fuller, J. F. C. (1943). *Armoured Warfare – An Annotated Edition of Fifteen Lectures on Operations Between Mechanized Forces*. London: Eyre & Spottiswoode.
Fuller, J. F. C. (1946). *Armament and History – A Study of the Influence of Armament on History from the Dawn of Classical Warfare to the Second World War*. London: Eyre & Spottiswoode.

Fuller, J. F. C. (1948). *The Second World War 1939–45. A Strategical and Tactical History*. London: Eyre & Spottiswoode.
Fuller, J. F. C. (1960). *The Generalship of Alexander the Great*. New Brunswick: Rutgers University Press.
Fuller, J. F. C. (1961). *The Conduct of War 1789–1961: A Study of the Impact of the French, Industrial, and Russian Revolutions on War and its Conduct*. London: Eyre & Spottiswoode.
Fuller, J. F. C. (1965). *Julius Caesar – Man, Soldier, and Tyrant*. London: Eyre & Spottiswoode.
Gat, A. (2001). *A History of Military Thought from the Enlightenment to the Cold War*. Oxford: Oxford University Press.
Gat, A. (2011). The Changing Character of War. In H. Strachan & S. Scheipers (Eds.), *The Changing Character of War* (pp. 27–47). Oxford: Oxford University Press.
Giap, V. N. (1968). *"Big Victory, Great Task" – North Viet-Nam's Minister of Defense Assesses the Course of the War*. New York: Frederick A. Praeger, Publishers.
Giap, V. N. (1970). *Banner of People's War, the Party's Military Line*. London: Pall Mall Press.
Glantz, D. M. (1991). *Soviet Military Operational Art - In Pursuit of Deep Battle*. Oxon: Frank Cass.
Goltz, Field-Marshal Baron von der. (1906). *The Nation in Arms – A Treatise on Modern Military Systems and the Conduct of War*. London: Hodder and Stoughton.
Grant, U. S. (2004). *Personal Memoirs of U.S. Grant*. Hazelton: A Penn State Electronic Classics Series Publication.
Gray, C. S. (2007). *Fighting Talk – Forty Maxims On War, Peace, and Strategy*. Westport: Praeger Security International.
Gray, C. S. (2010). *The Strategy Bridge - Theory for Practice*. Oxford: Oxford University Press.
Greenhalgh, E. (2011). *Foch in Command – The Forging of a First World War General*. Cambridge: Cambridge University Press.
Guibert, Jacques Antoine Hippolyte de. (1773). *Essai Général de Tactique*. Liege: C. Plomteux.
Handel, M. I. (2001). *Masters of War - Classical Strategic Thought.*. Third revised and expanded edition. London/New York: Routledge.
Heuser, B. (2002). *Reading Clausewitz*. New York: Random House.
Heuser, B. (2010). *The Evolution on Strategy – Thinking Strategy from Antiquity to the Present*. Cambridge: Cambridge University Press.
Howard, M. (1983). *Clausewitz*. Oxford: Oxford University Press.
Howard, M. (2008). *War and the Liberal Conscience*. London: Hurst & Company.

Isserson, G. S. (2013). *The Evolution of Operational Art.* (Bruce W. Menning, Trans). Fort Leavenworth: Combat Studies Institute Press.
Jomini, A. H. de. (1992). *The Art of War.* London: Greenhill Books.
Jomini, A. H. de. (2007). *The Art of War.* Rockville: Arc Manor.
Juvaini, 'Ala-ad-Din 'Ata-Malik. (1958). *The History of the World-Conqueror* (Vol. I). Cambridge: Harvard University Press.
Kagan, F. W. (2006). *Finding the Target - the Transformation of American Military Policy.* New York: Encounter Books.
Kaldor, M. (2007). *New & Old Wars* (2nd ed.). Stanford: Stanford University Press.
Keegan, J. (1998). *War and Our World The Reith Lectures 1998.* London: Hutchinson.
Kesselring, A. (1989). *The Memoirs of Field-Marshal Kesselring.* Novato: Presido.
Klein, J. J. (2006). *Space Warfare – Strategy, Principles and Policy.* London/New York: Routledge.
Lawrence, T. E. (1997). *Revolt in the Desert.* Hertfordshire: Wordsworth Edition.
Leo VI. (2010). *Taktika.* Text, Translation and Commentary by George T. Dennis. Washington, D.C: Dumbarton Oaks Texts XII.
Leonhard, R. R. (1991). *The Art of Maneuver - Maneuver-Warfare Theory and AirLand Battle.* Novato: Presido.
Leonhard, R. R. (1998). *The Principles of War for the Information Age.* New York: Ballantine Books.
Liddell Hart, B. H. (1927). *The Remaking of Modern Armies.* London: John Murray.
Liddell Hart, B. H. (1932). *When Britain Goes to War: Adaptability and Mobility.* London: Faber and Faber.
Liddell Hart, B. H. (1946). *The Revolution in Warfare.* London: Faber and Faber.
Liddell Hart, B. H. (1950). *Defence of the West – Some Riddles of War and Peace.* London: Cassell and Company.
Longworth, P. (1965). *The Art of Victory – The Life and Achievements of Generalissimo Suvorov 1729–1800.* London: Constable.
Lonsdale, D. (2007). Clausewitz and Information Warfare. *Clausewitz in the Twenty-First Century.* In H. Strachan & A. Herberg-Rothe (Eds.), (pp. 231–250). Oxford: Oxford University Press.
Ludendorff, E. (1919). *Sotamuistelmani 1914–1918.* Porvoo: WSOY.
Luttwak, E. N. (1987). *Strategy: The Logic of War and Peace.* Cambridge, MA: Belknap Press of Harvard University Press.
Luttwak, E. (1995), Towards Post-Heroic Warfare. In *Foreign Affairs* 74/3.
Machiavelli, N. (1965). *The Art of War.* A Revised Edition of the Ellis Farneworth Translation. Indianapolis: Bobbs-Merrill Company.
Mahan, A. T. (1999). The Influence of Sea Power Upon *History.* In *Roots of Strategy* (pp. 43–148, Book 4), Mechanicsburg: Stackpole Books.

Mandeles, M. D. (2005), The Future of War – Organizations as Weapons. Potomac Books Washington.

Mao, T.-T. (1963). *Selected Military Writings*. Peking: Foreign Languages Press.

Martel, G. (1945). *The Problem of Security*. London: Michael Joseph.

Martel, Giffard. Le Q. (1931), In the Wake of the Tank: The First Fifteen Years of Mechanization in the British Army. Sifton Praed, London.

Matheny, R. R. (2012). *Carrying the War to the Enemy – American Operational Art to 1945*. Norman: Oklahoma University Press.

McChrystal, S. (2013). *My Share of the Task*. London: Portfolio/Penguin.

Mearsheimer, J. J. (2010). *Liddell Hart and the Weight of History. Cornell Studies in Security Affairs*. Ithaca: Cornell University Press.

Mitchell, B. (1999). Winged Defence. In *Roots of Strategy* (pp. 409–516, Book 4). Mechanicsburg: Stackpole Books.

Moran, D. (2007). The instrument: Clausewitz on aims and objectives in war. *Clausewitz in the Twenty-First Century*. In H. Strachan & A. Herberg-Rothe (Eds.), (pp. 91–106). Oxford: Oxford University Press.

Münckler, H. (2007). Clausewitz and the privatization of war. *Clausewitz in the Twenty-First Century*. In H. Strachan & A. Herberg-Rothe (Eds.), (pp. 219–230). Oxford: Oxford University Press.

Musashi, M. (1995). *Maa, Vesi, Tuli, Tuuli ja Tyhjyys*. (Orig. *Go rin no sho*), Helsinki: Otava.

Napoleon Bonaparte. (1987). *The Military Maxims of Napoleon*. Ed. David G. Chandler. London: Greenhill Books.

Odom, W. E. (1993). *America's Military Revolution - Strategy and Structure after the Cold War*. Washington: American University Press.

Paret, P. (2009). *The Cognitive Challenge of War – Prussia 1806*. Princeton: Princeton University Press.

Patton, G. S., Jr. (1947). *War as I knew it*. Cambridge: Riverside Press.

Reid, B. H. (1998). *Studies in British Military Thought: Debates with Fuller and Liddell Hart*. Lincoln: University of Nebraska Press.

Reid, B. H. (2014). *The Origins of the American Civil War*. New York/London: Routledge.

Russell, B. (1959). *Common Sense and Nuclear Warfare*. London: George Allen & Unwin.

Schelling, T. C. (2008). *Arms and Influence. With a New Preface and Afterword*. New Haven/London: Yale University Press.

Scobell, A. (2011). The Chinese way of war. In J. A. Olsen & M. van Creveld (Eds.), *The Evolution of Operational Art – From Napoleon to the Present* (pp. 195–221). Oxford: Oxford university press.

Sherman, W. T. (1890). *Personal Memoirs of Gen'l W.T. Sherman* (Vol. I, 3rd ed.. Revised and Corrected. In Two Volumes). New York: Charles L. Webster.

Sherman, W. T. (1957). *Personal Memoirs of General W.T. Sherman* (Vol. II). Bloomington: Indiana University press.
Shy, J. (1976). *A People Numerous and Armed: Reflections on the Military Struggle for American Independence.* New York/Oxford: Oxford University Press.
Simpkin, R. E. (1985). *Race to the Swift – Thoughts on Twenty-First Century Warfare.* London: Brassey's Defence Publishers.
Simpkin, R. E. (1987). *Deep Battle - the Brainchild of Marshal Tukhachevskii.* London: Brassey's Defence Publishers.
Slessor, S. J. (1954). *Strategy for the West.* London: Cassell.
Smith, R. (2008). *The Utility of Force: The Art of War in the Modern World.* New York: Vintage Books.
Sokolovsky, V. D. (1963). *Military Strategy: Soviet Doctrine and Concepts.* London/Dunmow: Pall Mall Press.
Ssu-ma. (1993). The Methods of the Ssu-ma. In R. D. Sawyer (Ed.), *The Seven Military Classics of Ancient China* (pp. 107–144). Boulder: Westview Press.
St. Augustine. (1909), *The City of God (De Civitate Dei)* (John Healey, Trans., Vol. 2). Edinburgh: John Grant.
Stempel, J. (2012). *The Nature of War – Origins and Evolution of Violent Conflict.* Jefferson: McFarland.
Stoker, D. (2010). *The Grand Design – Strategy and the U.S. Civil War.* Oxford/New York: Oxford University Press.
Strachan, H. (2011a). Operational Art and Britain, 1909-2009. In J. A. Olsen & M. van Creveld (Eds.), *The Evolution of Operational Art – From Napoleon to the Present* (pp. 96–136). Oxford: Oxford University Press.
Strachan, H. (2011b). Strategy in the Twenty-First Century. In H. Strachan & S. Scheipers (Eds.), *The Changing Character of War* (pp. 503–523). Oxford: Oxford University Press.
Strachan, H. (2013). *The Direction of War – Contemporary Strategy in Historical Perspective.* Cambridge: Cambridge University Press.
Sun-tzu. (1993). Art of War. In R. D. Sawyer (Ed.), *The Seven Military Classics of Ancient China* (pp. 145–186). Boulder: Westview Press.
Svechin, A. A. (1992). *Strategy.* Boulder: Eastview Press.
Toffler, A. (1990a). *The Third Wave.* New York: Bantam Books, Paperback edition.
Toffler, A. (1990b). *Future Shock.* New York: Bantam Books, Paperback edition.
Toffler, A., & Toffler, H. (1995). *War and Anti-War: Survival at the Dawn of the 21st century.* New York: Warner Books.
van Creveld, M. (1991). *On Future War.* London: Brassey's.
van Creveld, M. (2005). *The Art of War - War and Military Thought.* New York: Harper Collins Publishers.
van Creveld, M. (2008). *The Culture of War.* New York: Ballantine Books.
Vego, M. (2009). *Joint Operational Warfare: Theory and Practice* (Revised Ed.). Newport: U.S. Naval War College.

von Bülow, D. H. (2013). *The Spirit of the Modern System of War*. Cambridge: Cambridge University Press.

von Clausewitz, C. (1989). *On War* (M. Howard & P. Paret, Ed. & Trans.). First paperback ed. Princeton: Princeton University Press.

von Freytag-Loringhoven, H. (1991), The Power of Personality in War. In *Roots of Strategy* (pp. 173–349, Book 3). Mechanicsburg: Stackpole Books.

von Moltke, H. (1993). *On the Art of War – Selected Writings*. Ed. D. J. Hughes. New York: Ballantine Books.

von Seeckt, H. (1930). *Valtiomies ja Sotapäällikkö*. Helsinki: Otava.

Warden, J. A., III. (1995, Spring) The Enemy as a System. *Airpower Journal*.

Warden, J. A., III. (2000). *The Air Campaign – Planning for Combat*. San Jose: to Excel.

Wavell, A. (1953). *Soldiers and Soldiering or Epithets of War*. London: Jonathan cape.

Wylie, J. C. (2014). *Military Strategy - A General Theory of Power Control*. Annapolis: Naval Institute Press.

CHAPTER 2

Manipulation of the Trinity of Time, Space, and Force

> The higher the level of war, the larger the factors of space, time, and force, and hence the more critical for the commanders and their staff to properly balance these factors with the respective military objective. (Vego 2009, p. III–3)

While Vego was right in claiming that space, time, and force are larger in the higher levels of war, the need to master and balance them is just as dire on every level. The quantities differ and this influences the respective meaning of the factors. In tactics, space is small, time is often short, and force is manageable. Tactics is about mastering the moment and acting in an instant. The balance is easier to create and favors quick timing and precise location. In grand strategy the entire world may be the space to be mastered with force created over a time span of decades. There may be no pressing hurry but the balance is precarious. Operational art is the tricky middle ground where there is simultaneously pressure to get the balance just right but depending on the situational needs some factor can be more important than other. This chapter seeks to understand why mastering the trinity of time, space, and force is so crucial; how one factor affects the others; and why in operational art time may be the most promising of them to manipulate.

J. Hanska, *War of Time*,
https://doi.org/10.1007/978-3-030-45517-0_2

2.1 Calculating Formulae of Victory

There are numerous attempts to treat the art of war scientifically. Simpkin (1985, p. xx) suggested a tripartite multidisciplinary approach consisting of physics, to understand the physical forces at play; statistics, since risk, chance, and surprise rely on it; and psychology, since the clash of two wills of the commanders is the decisive factor in war. Since chance affects both sides equally, "combat is as close to being deterministic as it is possible for any human activity to be" (Dupuy 1987, p. xxv). Maizeroy saw that while one part of war is mechanical and can be reduced from principles and taught by rules, the other is "quite sublime and residing solely in the head of the general, as depending on time, place and other circumstances, which are eternally varying, so as never to be twice the same in all respects" (Cited in Gat 2001, p. 42).

Montecuccoli attempted to reduce war-experience to universal and fundamental rules that could be applied to particular times and circumstances (van Creveld 2005, p. 75–78; Gat 2001, p. 22). Adhering to the Newtonian principles the indust-real science of war attempted to formulate laws instead of guidelines of the agrarian age. Analyses of the art of war through calculations of force-ratios have mostly been failures. None, however, have failed as spectacularly as Foch in his "mathematical demonstration of the truth" that with the level of sophistication rifles had reached before WW I their increased firepower would benefit the attack more than the defense. His calculations 'proved' an attacker can fire twice the number of rounds the defender could.

> As you see, the material superiority of fire quickly increases in favour of the attack as a result of improved firearms. How much more quickly will grow at the same time the ascendancy, the moral superiority of the assailant over the defender, of the crusher over the crushed. (Foch 1920, p. 32)

Fuller (1961, p. 124) claimed that while Foch was right to stress the importance of the offensive to his student, he erred in his willingness to "exalt it into a fetish." Many theorists and high-ranking officers have attempted to invent a formula for victory and some have even created equations for this purpose. A sad example is Dupuy (1987, p. 30) who summarized all of Clausewitz' philosophy on the art or war into a simple equation "$P = N x V x Q$". Combat power, P, for each of the opposing sides can be found by multiplying the numbers of troops, N, by variable circumstances affecting a force in battle, *V*, by quality of force, *Q*. His purpose was to create a calculus of combat and such formulae attempt to illustrate deep understanding of the mechanics of warfare. For Wavell

> all tactics since the earliest days have been based on evaluating an equation in which x=mobility, y=armour, and z=hitting power. Once a satisfactory solution has been found and a formula evolved, it tends to remain static until some thinking soldier (or possibly civilian) recognizes that the values of x, y, z have been changed by the progress of inventions since the last formula was accepted and that a new formula and new system of tactics are required. (Wavell 1953, p. 48–49)

This is another attempt to quantify war and reduce it into three factors without even an attempt to illustrate how their interrelations could affect the mechanics of war. Mobility, armor—or protection, and hitting power are indeed essentials of warfare, but they have to be combined in a unique manner specific to each occasion and Wavell doesn't instruct his readers how. Once someone discovers a suitable 'formula', it becomes the preferred method until cycle of war turns and forces re-evaluation. Then someone rewrites and revises it to better fit the changed context. This leads to new tactics and operational art and allows one to surprise his enemy. The task of a military genius is

> producing original combinations from the forces of war, genius must consequently be the mainspring of strategy, which is largely the science of forces. Inwardly its work is founded on originality; outwardly it manifests in surprise. The great genius surges through difficulties immune, because he sees—foresees—the end, and understands the means. (Fuller 1926, p. 99)

We can use Napoleon or Alexander the Great as examples of great captains who had an innate understanding of the science of war. Yet, Napoleon, for all his foresight, chose to attack Russia with inadequate resources. The reign of Alexander ended in a prosaic death. In both cases "though thousands have watched and followed him, to them his genius remains a mystery. The man is venerated, but his method vanishes, not because it is forgotten, but because it was never understood" (Fuller 1926, p. 98–99). There are too many attempts to recreate past deeds that seldom find the pearl of wisdom of what was the unique and exceptional thing that led to victory. Once found, the operational artist should adapt it to his context.

Napoleon argued that in war, "as in mechanics, time is the grand element between weight and force" (Cited in Fuller 1961, p. 50). As industreality was infatuated with technology, mass armies, and the use of factory-produced force in terms of weapons and mobility, physics and mechanics guided its formulae of war. One suitable equation is that of speed as distance divided by time. Another is that force equals mass multiplied by acceleration. Speed of the attack, depth of penetration, and the time consumed—their interrelations propose answers to many questions of an operational artist. If the force of the attack equals the mass of troops times their acceleration, to penetrate further into the opponent's defenses, the attacker needs more force. If he cannot increase the mass of troops he has to accelerate their movement, but warfare cannot be reduced to numbers. Speed, for example, can

> reduce space by economizing time. In other words, the more rapidly we move, the smaller becomes the bulk of the area we are called upon to defend. Strategically, time and space are relative, and as the history of war has shown again and again, a handful of men at a certain spot at a certain hour is frequently a far more powerful instrument of war than ten times the number on the same spot twenty-four hours later. If you can move five times as fast as your enemy, then, whilst the military hour will remain sixty minutes for him, it will be reduced to 12 minutes for you and every mile will become less than two furlongs. (Fuller 1943, p. 35)

For Fuller military time and space were relative. By increasing speed, time 'slows down' and in the same unit of time more distance can be covered. Likewise, vast spaces and distances can slow down the enemy and cause loss of time. This happened to both Napoleon and Hitler in attempting to conquer the vast emptiness of Russia. Time and space, speed and distance, force and acceleration became the fundamentals of theorizing operational art of the Second Wave. Thus,

> the strength of an army, like the quantity of motion in mechanics, is estimated by the mass multiplied by the velocity.' Armies to-day have become mass minus velocity—there is urgent need of research for the causes of this stagnant condition and for the speedy application of a remedy. (Liddell Hart 1927, p. v)

Liddell Hart was right to decry the lack of velocity. Speed and mass need to be balanced and increased effort was required to liberate warfare from its WW I paralysis. Unfortunately, no formula can be universal because war is waged between armies consisting of humans who add unpredictability into any equation. If wars abided to laws of mechanics, the superior in numbers would always win. Ingenuity and lack of intellect, willpower and weakness, passions and fears, all turn the simple formulae into insolvable ones. As Moltke (1993, p. 175) argued

> No calculation of space and time guarantees victory in this realm of chance, mistakes, and disappointments. Uncertainty and the danger of failure accompany every step toward the goal, which will not be attained if fate is completely unfavourable. In war everything is uncertain.

Uncertainty is increased by the human factor every individual combatant introduces to any equation. Operational planning must take into account that war is an interaction between opposing forces and during industrealty they were million-man forces. One's own operations are in a dialectic relationship with those of the enemy and both impact each other. While one attempts to increase his speed and slow down the enemy, the enemy acts alike, unless fighting is asymmetric to begin with. As Triandafillov (1994, p. 132) wrote, "the speed of an offensive, its pace, depends wholly on the frequency of the combat the attacker must conduct en route to the assigned target." The enemy will do his utmost to reduce speed and halt movement. This factor must be included into estimations of progress. In WW I, the Red Army had considerable pace and

> armies moved at a speed of 20–22 kilometers during the intervals between operations, but the pace fell to 7–10 kilometers per day when they had to move in contact. Naturally, it is difficult, it is simply impossible, to perform any kind of specific calculations on the pace of a particular front in a future war. (Triandafillov 1994, p. 135)

Thus, a calculation of pace is impossible to predict, since all depends on the enemy. Should the defense be effective, the attack may halt until a breakthrough is created somewhere. After the days of the great Suvorov who was hostile to stereotyped methods (Bazilevich 1945, p. 24), Russian tactics have been prone to mathematical calculations of ratios of forces and weapons. Nevertheless, in 1931 Triandafillov (1994, p. 165) claimed that,

> it would also be erroneous to look upon operational art as some sort of bookkeeping effort; it would be incorrect to convert operational decisions

> into simple arithmetic multiplication. [...] The art of the leader is to calculate the operational significance of these changing situational elements correctly and to determine the correct material and personnel resources required to accomplish a given specific mission.

Operational art is simultaneously a science and an artform without unbinding rules. All aspects of the science of war reflect human behavior in changing situations and Triandafillov emphasized this. The "situational elements" need to be calculated to reach the "specific mission". In war, every operation, every battle, every moment, is unique and can't be recreated in laboratory conditions.

However, just as an empirical scientist, the operational artist has to identify all external and internal factors that influence his experiment—the battle. He minimizes the possibility of unpredicted occurrences. For Clausewitz art was a creative and not an imitative or derivative activity. Warfare is governed by moral forces manifested in the clash of wills of the commanders (Howard 1983, p. 14, 30; van Creveld 2008, p. 307). Moral and material forces have to correlate properly in operational art. To rely on mass results in reduction of speed and wasted time.

> In an army endurance is intimately connected with numbers, and, paradoxical as it may seem, the greater the size of an army the more difficult it is to maintain its moral solidarity; for, as size reduces speed of movement, so does size reduce speed of thought and increases the area and speed of fear. (Fuller 1926, p. 313)

The logic partially abides to mechanics. Speed of movement is curtailed by mass, but mass acts as a multiplier of force. The bigger the army, the harder it is to maintain unified purpose and the more difficult its performance becomes. The commander alters the equation since his will acts as a catalyst of setting the mass in motion and his energy upholds momentum. Mass is a slow thinker but quick to panic and the commander must be just the opposite. For Ludendorff war was a struggle between unknown physical and mental forces and the weaker one's personality is, the harder his command in war becomes. The will of the commander is the only constant in war and everything revolves around it (Ludendorff 1919, p. 41). To manipulate time we should turn our backs to calculations and consider the commander and his mental abilities in conducting the symphony of war. It is his task to bind together scientific warfare and operational art.

He uses numbers and calculations but evaluates their validity in human interaction. We will return to this later when we discuss the mental aspects of winning time. For now, it suffices to say that there always has been a search for a universally valid formula of victory. However, as Beaufre (1965, p. 49) wrote, "War is a social phenomenon too complex to be governed by a single formula—unless it be so simple as to be a statement of the obvious."

2.2 Interrelations of Time, Place, and Force

> If I look upon the universe as space of three dimensions, then this space manifests to my mind in terms of time and force; time indicating the subjective relationships of mind and space, and force—the objective relationships. Time may be divided into past, present, and future; and force into energy, motion, and mass. We only know the past through the present, and can only speculate as regards the future from the present; and all our subjective knowledge in time is ultimately based on objective motion, or the relationship at any given moment between energy and mass. (Fuller 1926, p. 49)

Fuller as an early developer of operational thinking in his early works endowed the trinity of time, space, and force with almost mythic qualities, but these three factors are nowadays universally accepted as the foundation of efficient art of war. Each one is dependent on the others and the challenge of getting their balance to best suit the occasion is the main task of an operational artist. There have been several periods in military history when art of war was curtailed by societal conditions leading to negligence of time, space, and force as essential building blocks of any military masterpiece. As an example we can use the Gothic War in which progress was erratic. During this period Rome was conquered five times by Belisarius, Totila, and Narses in turn. The tactics and strategy were dictated by the fact that available forces were too small in relation to the vast area being fought for (Delbrück 1990b, p. 377; Procopius 1968). The interrelation of space, time, and force was crooked. At one point the Goths succumbed to the Byzantians after a four-year struggle without actual battles. Space in terms of distances and temporality through prolonged duration created a situation in which the insufficient mass of troops could not be used operationally or strategically to end the war. In a smaller area a decisive battle could have been created through denser troop concentration. Neither side was properly committed to the objective of ending the war rapidly. Thus, it was only when Justinian for the second time sent over a large army and the Goths simultaneously believed they had sufficient force to commit themselves that a decisive battle ended the war (Delbrück 1990b, p. 378–379).

Napoleon mastered creating favorable combinations of time, space, and force to attain his objectives and this is common to all successful commanders. The need to understand and alter the balance of these three factors is omnipresent as illustrated by Arthashastra. While many early writers concerned themselves only with finding the most promising moment for action, Kautilya (Book VII) attempted to teach how to recognize them. He gave specific directions when to prefer peace, when to remain neutral, and when to wage war. Even after the declaration of war, he specified the conditions and times suitable for actual engagements. Time was one of the crucial components of success in battle and a factor to be reckoned with. "*Strength, place, and time, each is helpful to the other*" (Kautilya, Book IX, Ch.1, p. 2). He considered the trinity to be an optimal combination of the three factors. If one is less suitable than the others, this affects the balance and likely outcome of the battle. One must choose with care the amount of troops he needs to amass, where to march them and when to engage the enemy. As Clausewitz (1989, p. 95) later put it,

> the whole of military activity must therefore relate directly or indirectly to the engagement. The end for which a soldier is recruited, clothed, armed, and trained, the whole object of his sleeping, eating, drinking, and marching is simply that he should fight at the right place and the right time.

Engagement should take place only when and where the commander chooses. Jomini was the first Occidental example of a strategist discussing the practical problems in the interrelation of space in time. He wrote that, "considering the difficulty of finding ground and time necessary to bring a very large force into action on the day of battle, an army of one hundred and thirty or one hundred and forty thousand men may easily resist a much larger force" (de Jomini 1992, p. 126). In Jomini's time logistics faced an enormous task of concentrating all the troops to the battlefield at the predetermined time. Thus smaller armies had a chance if the enemy failed to transport all his force to the specific place in time.

Assembling troops remained problematic until the American Civil War. Extensive use of railroads made Fuller (1961, p. 95) call it the first great conflict of the steam age. It became possible to concentrate far too much force on the battlefield to employ them. The armies at ends of railway lines swelled out of proportion and operability while remaining dependent on the railway. Local congestion was created and, "*the sum effect was to decrease the mobility of operations*" (Liddell Hart 1946, p. 12). Even if railroads made American Civil War a forerunner of operational mobility, they simultaneously hindered tactical movement. Because of the imbalance of firepower and tactics, after massacres such as the Second Bull Run the troops started to dig themselves into the ground for protection (Stempel 2012, p. 119).

The effects of congestion and entrenchment were multiplied if the defender managed to choose a battlefield with confined space for maneuver. Time and space limit each other and manipulating one can enable benefiting from the other. The idea of superior force striking at a decisive spot was the clue to Clausewitz's teachings. "To achieve strength at the decisive point depends on the strength of the army and on the skill with which this strength is employed. The first rule, therefore, should be: put the largest possible army into the field" (von Clausewitz 1989, p. 195). He also sardonically observed that the best strategy is to be strong first in general and then at the decisive point because superior forces provide the best chance for winning a battle (Howard 1983, p. 40). As Strachan (2007, p. 42) noted, "Space, terrain, and geography therefore contain the three elements of war—tactics, strategy, and policy—and they are at odds with time." For Clausewitz an army should always be employed with its maximum strength. Two armies on the battlefield are seldom equal and if there is no absolute superiority, force has to be concentrated to gain relative superiority for a given time at the decisive point. To succeed in this

> the calculation of space and time appears as the most essential factor, and this has given rise to the belief that in strategy space and time cover practically everything concerning the use of the forces. [...] But although the equation of time and space does underlie everything else, and is, so to speak, the daily bread of strategy, it is neither the most difficult nor the decisive factor. (von Clausewitz 1989, p. 197)

Winning a victory has nothing to do "with the ability to calculate the relationships of two such simple elements as time and space." The causes of victories lie in "energy for rapid movement, boldness for quick attacks, and the increased activity which danger generates in great men" (von Clausewitz 1989, p. 197). All these factors refer to the operational artist's ability to make time his ally. Instead of the simple decision of *this* time and *this* place, he needs to create complex equations of time and space. Clausewitz called for spatio-temporal compression of force.

> It is always true that the character of battle, like its name, is slaughter [Schlacht], and its price is blood. As a human being the commander will recoil from it. But the human spirit recoils even more from the idea of a decision brought about by a single blow. Here all action is compressed into a single point in time and space. Under these conditions a man may dimly feel that his powers cannot be developed and brought to bear in so short a period, that much would be gained if he could have more time even if there is no reason to suppose that this would work in his favor. All this is sheer illusion, yet not to be dismissed on that account. The very weakness that assails anyone who has to make an important decision may affect even more strongly a military commander who is called upon to decide a matter of such far-reaching consequences by a single blow. (von Clausewitz 1989, p. 259)

When reading Clausewitz one must always understand that in his writings he followed two parallel paths of inquiry and trains of thought. He made a distinction between the abstract 'ideal of war' and real warfare (Echevarria 2013). Both Moltke and Clausewitz argued that war should be concluded in victory as soon as possible. Compressing all action and lethal energy into the "single point in time and space" would guarantee that. Such an operation could vanquish the enemy with a single, decisive strike where and when it would have direst effects. It may result in momentary excessive bloodshed and slaughter, but would save lives by ending the war. This ideal of absolute war is often at odds with reality. For Bellamy (2007, p. 19) WW II on the Eastern Front was as close to absolute war as possible. For the soldier the ultimate goal in war is victory and attaining it as quickly as possible. "Whatever contributes to his quick victory usually diminishes overall casualties, especially his own. It is therefore self-evident to him that any device or tactic that hastens victory represents the highest morality" (Brodie 1973, p. 46).

Oriental thought includes the same attempt at rapid victory. The metaphor of water is common in the writings of Sun-Tzu. Usually it refers to armies moving in accordance with the terrain or filling the gaps left by the enemy, but in this case to time and force. To attain decisive victory "the combat of the victorious is like the sudden release of a pent-up torrent down a thousand-fathom gorge. This is the strategic disposition of force [*hsing*]" (Sun-tzu 1993, p. 164). Force for the decisive attack has to be collected over time and once the moment to act comes, the force must be unleashed like a torrent of water. Wei Liao-Tzu agreed with this idea. For him, "water is the softest and weakest of things, but whatever it collides with—such as hills and mounds—will be collapsed by it for no other reason than its nature is concentrated and its attack totally committed" (Wei Liao-Tzu 1993, p. 256–257).

This sudden rush of overwhelming force is evident in ancient Chinese art of war, but water has been used as a metaphor in the Occident as well. For example Clausewitz wrote how "the great event of battle would be like a mighty river sweeping away such a weak dike" (cited in Bassford et al. 2010, p. 82). We can use Clausewitz and treat the period of defensive waiting and rest prior to a battle as a state of equilibrium that covers both physical and psychological forces along with the circumstances. The period when either or both armies prepare to mount an offensive is when tension increases until movement in either direction commences. One gains and the other loses. The movement either will wear itself out or is halted by fresh troops of the defender and thus inactivity returns until a new cycle of tension and movement follows and direction of this movement may be contrary to the previous one (von Clausewitz 1989, p. 221). This distinction is, of course, highly theoretical, but enables us to glimpse at the dynamics of warfare in his time.

During the relaxation or equilibrium minor battles may occur, but no great decision is sought. The state of tension is of greater importance, since its effects are more decisive and great will-power and pressure are involved. Violence is about to be unleashed in an explosion-like burst of activity that will continue until its energy wanes (von Clausewitz 1989, p. 221). Energy, force, and time alike are compressed to be released in a sudden spasm. For maximized effect tension has to be built so that it can be triggered instantaneously. "War, in its higher forms, is not an infinite mass of minor events [...] War consists rather of single, great decisive actions, each of which needs to be handled individually" (von Clausewitz 1989, p. 153). Once the most advantageous time comes, the rapid and surprising attack commences. The Chinese art of war focused on waiting patiently for the moment to act and not being hasty. Time, place, and force need to find an equilibrium. The energy, force, and momentum of the troops are accumulated to be instantaneously released like a tsunami.

> One who excels at warfare will await events in the situation without making any movement. When he sees he can be victorious, he will arise; if he sees he

> cannot be victorious, he does not vacillate. Of the many harms that can beset an army, vacillation is the greatest. Of disasters that can befall an army, none surpasses doubt. One who excels in warfare will not lose an advantage when he perceives it or be doubtful when he meets the moment. One who loses an advantage or lags behind the time for action will, on the contrary, suffer from disaster. Thus the wise follow the time and do not lose an advantage; the skillful are decisive and have no doubts. For this reason when there is a sudden clap of lightning, there is not time to close the eyes. Advance as if suddenly startled; employ your troops as if deranged. Those who oppose you will be destroyed; those who come near will perish. Who can defend against such an attack? (T'ai Kung 1993, p. 69)

It is alien to our contemporary operational art to take an unhurried approach to time, awaiting the proper moment of action with a certainty that it will come. Such passivity goes against the deep-set need to seize the initiative. In the Chinese tradition, however, the general should "after deploying observe their actions. Watch the enemy and then initiate movement. If they are waiting [for our attack], then act accordingly. Do not drum the advance, but await the moment when their masses arise. If they attack, entrench your forces and observe them" (Ssu-ma 1993, p. 142). Timing the release of force was an intricate issue. It had to commence exactly when the advantage is greatest. But how to identify the moment to release the pent-up force with the most devastating impact? In Chinese strategic thought there are no fixed principles (Yuen 2014, p. 39), but there are situations when the enemy is most vulnerable to an attack;

> When the enemy has begun to assemble they can be attacked.
> When the men and horses have not yet been fed they can be attacked.
> When the seasonal or weather conditions are not advantageous to them they can be attacked.
> When they have not secured good terrain they can be attacked.
> When they are fleeing they can be attacked.
> When they are not vigilant they can be attacked.
> When they are tired and exhausted they can be attacked.
> When the general is absent from the officers and troops they can be attacked.
> When they are traversing long roads they can be attacked.
> When they are fording rivers they can be attacked.
> When the troops have not had any leisure time they can be attacked.
> When they encounter the difficulty of precipitous ravines or are on narrow roads they can be attacked.
> When their battle array is in disorder they can be attacked.
> When they are afraid they can be attacked. (T'ai Kung 1993, p. 96–97)

Against an enemy already settled in his defensive positions, one benefits from not hurrying. However, it was acknowledged in the Occident as well that "when the enemy is not yet deployed, taking the initiative is of the utmost value. In that case one must not hesitate to commence the battle, but one should always keep the battle objective consistent with available forces" (von Moltke 1993, p. 131). The ratio between the objective and available forces is crucial. If the enemy is unprepared for an attack, one can use less than the maximum obtainable force instead of waiting to amass troops. If the enemy is prepared to withstand an attack "everything available must be thrown into battle at all circumstances, for one can never have too much strength or too many chances for victory. The very last remaining battalion should therefore be brought to the battlefield" (von Moltke 1993, p. 128–129). Rapid action should be contemplated only in relation to the other factors: the preparedness of the enemy and the number of troops available. It was important for Moltke (1993, p. 131) "*to find the correct moment for commencing a serious engagement.*" In this he echoed Kautilya, who much earlier wrote how one should know the

> comparative strength and weakness of himself and of his enemy; and having ascertained the power, place, time, the time of marching and of recruiting the army, the consequences, the loss of men and money, and profits and danger, he should march with his full force; otherwise, he should keep quiet (Kautilya, book IX, ch. 1, p. 1).

As a general guideline to his offensive tactics it will suffice to say that "striking in all places and at all times, and striking by surprise are varieties of waging war with infantry" (Kautilya book X, ch. p. 9). This is wise advice if the commander has enough resources. If not, the principle of economy of force has to be adhered to. Brodie (1959, p. 26) noted that the meaning of economy has changed from its nineteenth century connotations of judicious management but not necessarily limited use. It is a violation of the principle to withhold force. If we look at the battles of Frederick the Great, he often was forced to fight with smaller forces than would have been preferable and sometimes even when he was tactically successful, he suffered higher casualties than his enemy (Bond 1998, p. 16).

We still attempt to use sufficient force at the right place at the right time, but the meaning of the time and place have drastically changed during the Third Wave. The classics of the agrarian age attempted to provide taxonomies of such suitable places and times (T'ai Kung 1993, pp. 96–97, 102–103; or Kautilya, Book X, ch. 3, p. 3–4). Latter theorists omit such guidance since operational art has become more complicated as dimensions of space have become less important. Distance between the operational artist and his theater has diminished in relevance because

> information, and the decisions that result, can travel almost instantaneously to the place(s) where they are needed, making the location of those who gather, analyze, make decisions, and possibly those who act on these decisions, largely irrelevant. The Information Age is also compressing the time dimension. First, by making location less important, it reduces the need for time-consuming travel, whether local or long distance. (Alberts et al. 2000, p. 20)

This is a delicate issue and it must be understood that the mobility and location of the troops as well as their headquarters, and, to a smaller degree, even weapon platforms are losing importance. It all comes down to a new adaptation of the idea not to send a man if you can send a projectile. *Forces* don't have to be spatiotemporally concentrated, but their *effects* must. Information travels without human messengers and projectiles possess increased range and precision. Few armies have reached the stage of technological sophistication required for the kind of warfare network-centric age theoretically would enable. Thus, the demand for traditional mobility isn't going to vanish any time soon.

2.3 Conquering Space to Win Time

Fuller spent a lot of time perusing the interrelations of space, time, and force and he perceived them as highly specialized concepts in the art of war. In other words, he saw their meaning and essence to differ completely between the times of peace and war.

> In war these three conditions surround us as completely as they do in peace, but as our minds are concentrated on a single and highly specialized problem, namely the waging of a war, they assume a relatively military aspect, and, in order to distinguish them from their more general forms, I will call them military space, military force, and military time. (Fuller 1926, p. 175)

This separation is artificial, but space, time, and force do mean something different in war than their general description entails. Physically they are inflexible constants but in warfare they are dependent on human actions and the interplay of the enemies. What one side does affects the other. Time, space, and force become relative and the way belligerents experience them shapes them and gives them meaning. Time in warfare is not measured only in seconds and minutes but also by what the belligerents can accomplish during those seconds and minutes. Fuller (1926, p. 180) wrote that the "practical application of time is the utilization of space."

Theorists use different explanations for successes of certain armies and commanders. Liddell Hart was fascinated with mobility and he sought for reasons for success in how fast the troops moved. According to his interpretation, Roman legions maintained dominance due "to the development of a network of strategic roads. Science did not enable the Romans to produce mechanical legs, but they gave their soldiers roads on which they could use their legs to the best advantage" (Liddell Hart 1932, p. 65). Delbrück (1990a) claimed that the success of the Romans was explained by their ability to organize the chaos of war, starting from the battle formations. Both interpretations are correct, since organization allowed the troops to march faster and roads made it possible to estimate better how far they could proceed in a day. The Roman road network was a miracle of the Ancient world from the perspective of strategy and operational art. All roads led to Rome. At their other ends the roads led to the remotest corners of the empire and were used to send out the legions to pacify and protect the empire's frontiers. Roman road network enabled commerce but also rapid concentration of military force. In a hand-to-hand melee, with equal weaponry, a legionnaire might have succumbed to a barbarian enemy. Superior tactics and road network organizing the strategical or operational level movements of the legions made the Roman battle system more efficient and allowed saving time both in preparation and execution of the battle.

In early days of indust-real warfare from Moltke to WW I study of railroad maps provided understanding of possibilities to transport troops and guidance where battles were likely to occur. Railways were of immense importance in strategy. In 1870 Moltke "predicted exactly where the French army would concentrate. […] In order to fathom this state secret he limited himself to the cost of a simple railway map" (Schlieffen 1936, p. 64). He also developed a method for extensive use of the railways for offensive purposes. Railways enabled quick operational transport of troops.

Matheny (2012, p. 12) noted that the key element of Moltke's operations was rapid mobilization coupled with concentration of force and he exploited new technologies. For the first time in history an entire mobilization plan including timings and sequences was laid out beforehand and military needs surpassed all other rail traffic (Bucholz 2001, p. 51).

Even if railways had a profound impact on time in operational art they didn't eradicate its meaning. Theoretical calculation of rail-speeds and capacities of trains doesn't equal time consumed in transportation of troops. Preparations must be made, orders issued, and troops marched to the railway stations. On some rail lines trains could be sent only at given intervals and transportation with necessary pauses consumes time. This led Ludendorff (1919, p. 410) to estimate that transporting a division required approximately thirty trains and thus it took two or three days to move it to the Western Front. By the end of the nineteenth century railways had become a necessity to mobilization and maneuver, but instead of improvisation they required precise calculations of timings and capacities with adherence to timetables (Vego 2009, p. I–17). Moltke's operational art

> exploited the railway in order to give greater tactical and strategical mobility. Napoleon, because of a lack of roads and poor intelligence, had been forced to bring masses of men to the decisive point for a concentrated blow—interior lines, in fact. Moltke, with the new means of transportation and communication (the railway), developed a new method—a scattered deployment on exterior lines. (Montgomery 1961, p.32)

Operationally exploitation of the railway hinged on maximized utilization of the entire railway network to minimize the transport time. When both sides began to deploy their troops to the vicinity of the border, "in this race the timing is no longer by day, but by hours. It is therefore, of the highest importance for strategical concentration to use as many railways as possible, if possible all leading in the direction of the theater of war" (Goltz 1906, p. 97–98). Yet operationally on distances shorter than hundred kilometers for division-sized troops it was faster to march. Furthermore, once rail-deployment of an army had started, the time-table couldn't be altered (van Creveld 1991, p. 107).

> Armies came to depend on the railway for their maintenance without fully realizing how dependent they had become. Increased ease of supply encour-

> aged them to swell their numbers—at the end of the railway line—without asking themselves what effect those numbers would have on their power of action. (Liddell Hart 1932, p. 62)

In mid-seventeenth century there was a system of five-day marches that effectively meant that armies could go no further than 125 miles from their base of operations (von Bülow 2013). The same system was resurrected with the railway depot taking the place of a base tying troops to railroad junctions (Svechin 1992, p. 258–259). Even during WW I it was accepted as an established fact that to receive logistic support an army could not advance further than 150 kilometers from its railhead (von Manstein 1982, p. 411). As von Schlieffen (1936, p. 64) put it, "in the age of railways, the advance of each army is dependent upon and determined by the railway net. It may be laid a little forward or a little backward. But in the main it is fixed. Even today this is true although a dense network of railway lines covers the country." Rail transportation was the most economical means when distances to cover and the quantities of material and manpower to be moved were extensive.

Even if railways became the preferred method of strategic locomotion during the second half of the nineteenth century, railway is an inflexible instrument. Troops could only go where the rails went (van Creveld 1991, p. 107). In the mechanized and motorized era roads became the focal point again. As the number of automobiles increased the road network started to supplement operational movement and during WW I trucks began to infiltrate the battlefields as means of troop transport. They increased tactical mobility by enabling movement of troops and weapons from one sector of the battlefield to another in what Luttwak (1987, p. 158–159) calls "*tactical time*", that is, during the course of a single battle. Tanks and trucks benefited from the road network and remained dependent on it for operational movement as well. This was a return to First Wave warfare when roads were essential for widespread empires. Darius could create the Persian Empire and assure speedy transport of necessary troops to quell unrest by turning caravan routes into roads suitable for his armies (Fuller 1960, p. 76). George Patton Jr. (1947, p. 92) jotted down in his diary, "*surely the greatest study of war is the road net.*" Road maps gained a prominent position in planning the operations of motorized warfare during the indust-reality.

> In the High Command, small-scale maps are the best because from that level one has to decide on general policies and determine the places, usually road centers or river lines, the capture of which will hurt the enemy most. How these places are captured is a matter for the lower echelons to determine from the study of large-scale maps or, better still, from the ground. (Patton 1947, p. 92)

The higher up one is in the military hierarchy, the bigger the picture one must deal with. Thus one concentrates on sea, rail, and road routes as well as harbors and airfields. In tactics, it is important to grasp implications of minuscule details of terrain. A strategist concerns himself with abstractions and deals mostly with dimensions. In operational art the abstract principles and hard facts of infrastructure and geography intermingle. As Fuller argued,

> formerly space, from its military aspect, was two dimensional as regards tactics and one dimensional as regards supply. The addition of a second dimension to supply, by means of the cross-country tractor, and of a third dimension to tactics, by means of the aeroplane, both petrol driven machines, has ushered in a new military epoch. (Fuller 1926, p. 180–181)

But this new epoch of mechanization only showed the way to proceed further. One example how strategical troop movements have become rapider is that at the beginning of Operation Desert Storm there were as many troops in the Gulf region as in Vietnam at the height of the war. The troop build-up in Vietnam took four years but in the Gulf it was accomplished in six months. In the end every airfield in the region was so packed with aircraft that offers for more had to be declined (Schwarzkopf 1993, p. 403, 456).

> Time, strategically, is the measurement of military movement; tactically, of muscular and mechanical endurance. Time is, therefore, intimately related to the means of movement, protection, and weapons. [...] Time, also, frequently means concentration and economy of force. Thus, if time can be economized, numbers can either be multiplied or reduced, especially if an operation is carried out so rapidly that the enemy is unable to meet it. Superiority of time is so important a factor in war that frequently becomes the governing condition. (Fuller 1926, p. 180)

Time is entwined with force on a fundamental level, and operational art requires grasp of this relationship. Economization of time can create and dissolve force concentrations. Manipulation of time can act as a force multiplier. An attacking force may be able to strike and withdraw with minimal casualties if it performs faster than the defender. Nevertheless, Montgomery, one of the great economizers of force, reduced the speed of his troops. This shows that in the long run economy of force and velocity in warfare are not compatible (Reid 1997, p. 100). A choice has to be made regarding what to economize. Time can be saved or squandered depending upon the requirements of the situation, but to attain superiority over the enemy the skill to manipulate time may be the key.

> Time is the controlling factor in war, it is the urge of armies. The more rapid the assembly, the quicker the battle; the quicker the battle, the more speedy the victory, and victory is the postern of peace. Petrol economizes time in war, caterpillar tracks or aircraft propellers economize space. Economy in time and space are the sire and dam of surprise, and surprise is the true sword of victory. (Fuller 1923, p. 171)

Operationally time and space are tightly connected to the will of commanders and the clash of their forces. The indirect approach, the German *Blitzkrieg*, and Soviet deep operations relied on locating the weak points of the enemy and racing into the depth as far and fast as possible. MacArthur (1964, p. 57) described it as "tactic of breaking through and then by-passing strong points to exploit the lightly held rear areas." Yet only the speed of mechanized forces allowed for truly deep penetrations. One of the shortest and simplest definitions of mechanized tactics and operational art came from Rommel (2006, p. 165), who wrote that, "it is vital that the leading elements penetrate a maximum distance in a minimum time and that there be no diversification of effort in smashing into the first positions." All important elements are included. With speed, time is converted to distance. As far as possible, as soon as possible was the recipe of mechanized attack.

Indust-reality brought a profound change in the meaning of distances. Petrol engine, the tank, radio communication, and military aviation were changes in the ways and means of warfare that increased speed while distances could be grown and yet controlled. The dawn of the information society had the same effect, only more drastically. "Economies of speed replace economies of scale. Competition is so intense and the speeds required so high, that the "old 'time is money' rule is increasingly updated to 'every interval of time is worth more than the one before it'" (Toffler and Toffler 1995, p. 63). This illustrates the intellectual switch from distances to times spent in crossing them. Speed conquers scale. Time dominates distance and with each technical innovation speed keeps accelerating. Super-sonic weapons and planes are commonplace today. Velocity in information and cyber domains is limited practically only by the speed of light. Time is still money, like during indust-reality, but perhaps today nanoseconds are worth Bitcoins. The old adages are true, but their meanings change. Even space has changed in the course of the evolution of war.

2.4 Growth of Battlespace

The more sophisticated warfare has become, the further warriors have retreated from the frontline, and the distances between the combatants have grown thus increasing the size of the battlefield. In the old days, armies amassed for battle so that infantry, cavalry, and artillery alike were pushed to the very front to maximize their effect. Before long-range weapons "battlefields were of about the size of the present manoeuvring ground for a brigade. Even those who visit the battlefields of 1864 are astonished to find how short all the distances are" (Goltz 1906, p. 34). Prior to the World Wars Bloch (1914, p. 5) was able to correctly predict several development patterns that came to characterize indust-real and mechanized warfare. These were "(1) The opening of battles from much greater distances than formerly; (2) the necessity of loose formation in attack ; (3) the strengthening of the defence ; (4) the increase in the area of the battlefield; and (5) the increase in casualties." There were two preconditions to growth of the size of the battlefield and dispersion of forces within it: mobility of troops and the range of weapons. The more troops one could transport into the battle in time, the larger the area these troops occupy. But it would be useless to concentrate excessive force unless weapons permit every soldier to do his part.

When swords, halberds, and lances were the primary weapons, the troops could and should be in compact formations since they would engage the enemy in hand-to-hand combat. There was no need for extended battlefronts. As Isserson (2013, p. 16) put it, even in the Napoleonic era battle "*possessed no spatial dimension because its scale consisted of a single point, and it had no temporal dimension because it was simply a moment in time. Moreover, it had no depth because it took place in a locale.*" In classical warfare battles seldom lasted for more than a single day (Bond 1998, p. 1). Since troops clashed physically, a relatively minor area was sufficient for battle. To some extent even a nineteenth century battle was composed of

> a series of clashes taking place over a short period of time in a small area with the enemies in close proximity; the total duration of a battle was not too much greater than the duration on an individual clash (...) but conditions began to change once the Russo-Japanese War began. The battle front began to get larger, and the sites of battles became fragmented and were great distances apart from one another. (Svechin 1992, p. 271)

Expansion of battlefields started gradually during the seventeenth and eighteenth centuries with the introduction of advanced guns and rifles and conscription (Vego 2009, p. IV–3, 9; Colby 1943, p. 17–20). Firearms enabled engaging the enemy at a greater distance and the development in means of transport made it possible to assemble enormous forces to the battlefield (du Picq 1987, p. 128). At the same time the intensity of fire enforced a loosening of infantry fighting formations. The point of battle turned into a linear front and thus with the same amount of troops, the battlefields grew "in extent out of proportion with the troops engaged" (von Bernhardi 1914, p. 62). Growth in the physical dimensions of the battlefield grew its temporal dimension as a by-product. "In the old days the time scale was small (a nineteenth century campaign might last a month, a battle a few hours); in the great wars of the twentieth century the time scale was longer" (Beaufre 1965, p. 100–101). While Frederickian and Napoleonic battles were bloodier than others fought before the World Wars, there was nevertheless a profound difference between these types of battles. Namely that earlier the lack and ineffectiveness of long-range weapons "made the moments of maximum danger shorter. In our time the fire may last for hours at long range" (von Freytag-Loringhoven 1991, p. 211). Frederick wanted to engage the enemy rapidly in close combat. For this reason, he drilled his troops to march with a quick step and rifle on the shoulder until the melee.

> It is not the greater or lesser number of dead that decides an action but the ground you gain. It is not fire but bearing which defeats the enemy. And because the decision is gained more quickly by always marching against the enemy than by amusing yourself firing, the sooner a battle is decided, the fewer men are lost in it. (Frederick 1985, p. 396)

This type of tactical thinking became obsolete as rifles became rapid to reload. Clausewitz lamented the fact that "our generals are too taken with the idea that it is better to advance than to stand and fire. Each of these actions has its proper place" (Cited in Bassford et al. 2010, p. 123). During the latter half of the nineteenth century it had become evident that rifled and breech-loaded firearms necessitated a revision of infantry tactics. These preliminary findings were justified by the experiences of the Boer and Russo-Japanese Wars, but only during WW I battles like Somme and Passchendaele testified to the fact that infantrymen can indeed do more than "amuse themselves" firing (Echevarria 2000, p. 13–14; Bond 1998, p. 99). Modernization of firearms had effects beyond existing tactics. Since they had a direct influence on combat and therefore conduct of operations, "we have had to acknowledge that, indirectly, strategy is affected as well by the altered nature of battle" (von Bernhardi 1914, p. 71).

Before efficient rifles two armies could closely observe each other from a relatively close distance. When the decision to attack was made, the terrain in-between was crossed by foot soldiers. Delbrück (1990a, p. 76) noted that neither a Greek phalanx nor any other closed formation was able to run a mile in any case. In other words, ancient armies moved with a relatively sedate place toward the enemy until they reached the range of arrows. The time spent in closing in on the enemy was short and the enemy could not inflict damage with his weapons during the whole duration of that time. Long-distance weapons changed that. First the rifle prolonged the time the attacker had to spend under fire in advance, and with increased artillery range even the enemy supply units, other auxiliary troops, and reserves could be engaged. Better accuracy, range, and rate of fire pushed the infantry soldiers to their psychological breaking points and infantry tactics had to be revised to make attacking possible at all (Echevarria 2000, p. 213–214). The fascination of inflicting casualties from a distance was soon lost due to lack of protection of one's own troops since the enemy had the same capability. As despite the physical distance casualties still occurred constantly, there was a need to rapidly close the gap again.

> Every soldier should realize that casualties in battle are the result of two factors: first, effective enemy fire, and, second, the time during which the soldier is exposed to that fire. The enemy's effectiveness in fire is reduced by your fire or night attacks. The time you are exposed is reduced by the rapidity of your advance. (Patton 1947, p. 340)

Foot soldiers remained exposed to enemy fire for too long, and faster means of crossing the distance were needed. The tank was a relatively slow method of transportation to begin with but ultimately provided the answer. In the initial stages tanks benefited from being impregnable by rifle fire and slowly crawling across the gap. As anti-tank weapons developed, the tank had to gain speed. Movement equals protection, since one being a target remains for a shorter period of time. Man or machine, the principle is the same, and to reduce casualties the duration had to be cut by accelerating the pace of crossing, shortening the span of no-man's land, or seeking some other form of protection. "The art, then, is to reduce this zone of advance, in launching the attack from as short a distance as possible. Ground provides the means for it" (Foch 1920, p. 350). Features of the terrain could be used as protection from direct fire to get close to the enemy. The destructive potential of machine guns made crossing open swathes of ground hazardous and the losses suffered were extensive. Tactics had to negate the effect of enemy fire on advancing troops and still shorten the time spent under fire. Time is the most important commodity in warfare. The right moment and the right speed of action will save lives and perhaps even enable a victory in a hopeless situation, because "minutes are momentous, and the history of war has countless proofs that a handful of men can often gain a position which half an hour later a thousand cannot gain, while half a day later 10, 000 are too few" (Liddell Hart 1932, p. 193–194).

Developments in propulsion of artillery projectiles kept increasing the range and thus growing the battlefield in depth, but dispersion of troops influenced the width as well. The bigger the two-dimensional battlefield grew and the more extended the fronts stretched, the more important mobility became. Fuller (1946, p. 23) wrote that "as reach of range is the dominant characteristic of an offensive weapon, speed and mobility in attack are the dominant characteristics of the offensive itself." This has been a constant development pattern throughout history. As weapons have become more lethal, their targets have spread out along the battlefield to minimize casualties (Dupuy 1987, p. 170–172). To gain protection troops on operational level needed to disperse or be pulled back beyond the range of artillery. As the distance grew one had to invent faster means of transport to turn troop dispersion into troop concentration and achieving local material and manpower superiority over the enemy at the decisive points of an operation. Difference between the World Wars was one

> of space and speed rather than of weapons. The weapon development favourable to the offensive has been counter-balanced by the weapon development favourable to the defensive. But other conditions have made the present war less static. While defence is stronger, space has been wider and forces faster. These offsets have given the offensive a better chance strategically than in the last war—by providing the attacker with more room for manoeuvre, and more speed of manoeuvre, thus making it easier for him to achieve penetration, so long as he avoids pushing into bottlenecks. (Liddell Hart 1946, 22–23)

Even if defensive formations were stronger, the mobility of the attacker weakened the defensive by forcing it out of concentration to cover a wider area. Defense had to disperse as well to counter the agile and mobile attackers. Between the World Wars static trenches extending across the entire continent turned into wide maneuvers and forces dispersed prior to the battle to be concentrated locally and temporally for the attack. Still it remained true that the

> enemy must be found, held and hit, whereas the finder must strive his utmost not to be found, not to be held and not to be hit. Yet every weapon is influenced not only by ground, time and space, but by every other weapon—the enemy's and our own. No new weapon can be introduced without changing conditions, and every change in conditions will demand a modification in the application of the principles of war. Once the enemy is discovered, the whole theory of the attack must be based on a careful study of ground with reference to offensive action and protective power, as well as the time it will take both sides to move over the ground. Consequently, correct timing of movements becomes the decisive factor. (Fuller 1943, p. 85)

Again, the application of principles of war requires modification as a response to time, space, and weapons specific to the situation. Time is a fundamental factor in these calculations: time required to cross the distance, time when the offensive needs to hit the target, time as the factor controlling and synchronizing movement, and time as a resource needed to understand and execute orders. To find, hold, and hit the enemy presupposes the ability to spatio-temporally locate, immobilize, and influence it. Air power was an important asset in this.

Highlighting its mobility and speed, Liddell Hart (1932, p. 99) eloquently wrote that the "aircraft came endowed with a knight's move to supplement the military bishops and rooks on the chessboard of war."

Almost as if on a chessboard, the primitive battles focused on one point and after the king fell the fight was over. As armies improved their organization, battles were fought in linear formations and mechanized troops brought about area tactics making wars two-dimensional. Air power enabled going over the rigid lines and attacking rear areas in depth. It forced thinkers to readdress the applicability of old theories. War now took place in the third dimension as well, in effect turning the battlefield into a battlespace (Freedman 2006, p. 14). Air power made all Europe battlespace in WW II since bombers from England could drop their payloads on German cities.

> By virtue of this new weapon, the repercussions of war are no longer limited by the farthest artillery range of surface guns, but can be directly felt for hundreds and hundreds of miles over all the lands and seas of nations at war. No longer can areas exist in which life can be lived in safety and tranquility, nor can the battlefield any longer be limited to actual combatants. On the contrary, the battlefield will be limited only by the boundaries of the nations at war, and all of their citizens will become combatants, since all of them will be exposed to the aerial offensives of the enemy. (Douhet 1999, p. 283)

Next turning point in mobile warfare and the use of the aerial dimension was reached in Vietnam War and more specifically the Ia Drang Valley Campaign in which helicopters were used for the first time to maneuver large U.S. units in battle. As such, Schwarzkopf (1993, p. 140) claimed," it was a landmark in modern warfare." Helicopters added to tactical and operational mobility through air mechanization. Increase in mobility seeped even into the strategical level as maneuvers for the troops that could be airlifted became inter-theater and performed in operational time. Rotary-wing aircraft supplemented the development of the use of air space to the degree that since WW II battles have no longer been fought over areas but

> in cubic spaces. Consequently the battlefield of to-day may be compared to a box in which the armies contained in it, whether stationary or moving, are, or at least should be, constantly prepared to defend themselves on all sides—top, front, rear and flanks—or assume the offensive in one or more of these directions. (Fuller 1948, p. 46)

Mechanized units, their operations closely coordinated with those of the air force, and increased rotary-wing air-mechanization of units made operational movement of troops around entire theaters faster than ever. In a given theater of war air and naval forces interact with ground forces and since their dispositions can be altered swiftly, the spatial aspect of strategic war has become less important. Still, on operational and tactical levels momentary location of troops can be decisive (Luttwak 1987, p. 157). Everything depends on maneuver since it further diminishes the meaning of space. Space has in operational sense been overcome by time.

Air power has been succeeded though not superseded by space and cyber power (Gray 2012, p. 57). As we move toward operations in space, information domain, and the cyber domain, we are approaching a time when time and space don't correlate at all. However, military history teaches that such a profound revolution is highly unlikely. The impact of new technology is overestimated. When joint operations are planned and synchronized, the methods and weapons of cyber warfare only add options to operational artists. Nevertheless our information age battlespace has stretched to cover new dimensions and it is today more of a mental construct than a physical realm. In the first Gulf War

> The front was no longer where the main battle occurred. Precisely as called for in AirLand Battle doctrine, the allies were deepening the battle in all dimensions—distance, altitude, time. The front was now in the rear, at the sides, and up above. Actions were planned twelve, twenty-four, seventy-two hours ahead, choreographed in time, as it were. (Toffler and Toffler 1995, p. 68)

In Iraq the U.S.-led forces had absolute air superiority and dominated the sky of the battlespace. The 'front' was established at such a high altitude that the Iraqi troops could not contest it. The most important thing, however, was extension of the battlespace into the fourth, the temporal, dimension. Actions were choreographed ahead for pre-set periods. This meant that the battlespace was no longer only here and *now*, but included elements of the future as well. Battlespace had expanded from the physical dimensions to intangible ones as well. Even information itself has become a battlespace per se (Alberts et al. 2000, p. 60; Dillon-Reid 2009, p. 111).

But what does this actually mean and imply? Napoleon envisioned future circumstances after each action to win time from the enemy by being one step ahead. Yet, there is a world of difference. Napoleon thought of things that could happen, the planners of the Third Wave what will happen. Actions were determined in advance. They would happen at pre-set times and all activities were synchronized together. Thus, the battlespace stretched from the present moment into the future and was no longer bound by the concept of here and now. This required new ways of thought and increased the demand for mental flexibility. Synchronization of future activities doesn't contradict the idea that the fog of war makes it impossible to see far ahead, or, in other words "no plan of operations can with any safety include more than the first collision with the enemy's main force" (Goltz 1906, p. 105). No one assigns synchronized tasks to the ground forces beyond the battle they are fighting, but timings of certain effects could be predetermined. An air strike can be ordered to a specific place at a pre-determined time and as soon as the order was processed, one could conceive of the attack as an eventuality. This can be used to set the rhythm the rest of the instruments of war could be synchronized with.

This conceptual spread into the temporal dimension meant that wars can no longer be confined to physical geography. When traditionally one state declared war against the other the border between two was the primary locus of dispute. The front shifted after attacks and defensive maneuvers if either side made progress or remained relatively stable. "It is no longer possible to contain war geographically. Zones of peace and zones of war exist side by side in the same territorial space" (Kaldor 2007, p. 178–179). War is simultaneously everywhere or nowhere. The tangible battlefields of the agrarian age and the indust-reality have been replaced by an ambiguous concept of multi-dimensional battlespace during the Third Wave. The wide variety of operations a great power can conduct makes battlespace "fluid." As Vego (2009, p. IV–9) describes them

> battlefield and battlespace are "fluid" or "transitory" parts of a theater. Their physical extent can vary greatly, depending on the scale and number of the objectives to be accomplished, the size and force mix on both sides, and the characteristics of the physical environment in which combat takes place. The battlefield and battlespace are not part of the permanent or semi-permanent theater structure. Their size in all dimensions varies greatly and is not constant; it can be expanded or contracted depending on the dynamics of combat. They are also not fixed in space, but movable. (…) In generic terms, "battlespace" pertains to a three-dimensional physical space, plus a corresponding part of cyberspace, in which a tactical commander sets the terms of a battle or engagement. Normally, a battlespace is larger than an area of operations, and it can overlap the battlespace of other friendly commanders. It expands and contracts depending on the dynamics of the battle or engagement and the capabilities of forces on both sides.

While battlespace is most commonly used to refer to the aforementioned three dimensions and cyberspace, it is more feasible to refer to a spatio-temporally bounded but fluid locus in which combat occurs at a given time and acknowledge that this space contains three of more dimensions. Even before the cyberwar hubris caught on there was a similar hype about the information domain as the fourth dimension of battlespace. Franks (2004, p. 175) argued that only digitalization turned battlefield into battlespace. But physicists tend to agree that time is the fourth dimension. Indeed, Waterloo, Cannae, or any site of past battles were battlespaces only for the duration of the combat. Time is one of the dimensions and defining characteristics of the battlespace. Temporal mastery of the battlespace is demanded from the operational artist. As Smith (2008, p. 21) wrote about Waterloo, "if the battle had taken place a month later it is possible all these factors (and the outcome) would have been different—this dependence on the circumstances of the day and understanding their significance is the true framework of military activity."

Nevertheless, Vego provided us with an important definition. Since the battlespace has extended in all three physical and more than a few intangible dimensions beyond the grasp of Scipios or Hannibals of the past, the contemporary concept has to be approached through a flexible and fluid perspective. Battlespace exists only for the duration of the fighting therein and its size can vastly vary depending on the operations conducted at any given time. Just as many other things in today's warfare are not stable and fixed, the idea of battlespace is fluid and transitory as well.

2.5 Trading Space to Time

> On land, large space provides the defender sufficient room for maneuver, both laterally and in the depth of one's rear. Large space allows the defender to trade space for time, if necessary. (Vego 2009, p. III–8)

As Vego noted, if space is available, it can be traded to time. However, as time is always lacking in warfare, it the equation that does not work both ways. The importance of time compared to terrain was recognized in the beginning of mechanized warfare. Martel (1931, p. 57) wrote that "in all economies the most important factor is the economy of time. The tank was pre-eminently a timesaving machine in France." He was a young officer at the time of his infatuation with the promise of the tank. A true child of the industrial age, he viewed time as a commodity and with the tank, less space could be exchanged into longer periods of time. This quotation was written long before the sweeping maneuvers of WW II at a time when tanks crawled onward at merely a few miles per hour. Once speeds accelerated, more time could be saved with mechanized forces by gaining or giving up space for other operational purposes. When it comes to the relationship between time and space in warfare Gneisenau earlier summarized the essence of managing and balancing these two in a few weighty words. "Strategy is the art of utilising time and space. I am more economic of the first than of the second. I can always regain space; time lost, never" (Cited in Foch 1920, p. 81). Thus one can

> regard a voluntary withdrawal to the interior of the country as a special form of indirect resistance—a form that destroys the enemy not so much by the sword as by his own exertions. Either no major battle is planned, or else it will be assumed to take place so late that the enemy's strength has already been sapped considerably. (von Clausewitz 1989, p. 469)

The lines of communication and supply may become so elongated that the enemy offensive wanes of its own accord. In First Wave agrarian warfare the most important cities of Europe were strongly fortified, but the countryside was available for such exchanges of space to exert the enemy. During indust-real mechanized warfare, the deserts of North Africa offered the same option to Rommel. Even if one cedes ground to the enemy, it can be won back later if the campaign proceeds as planned. Space is there for re-gaining, but whatever time one loses to the enemy is lost forever. The more important the operation, the less consequential becomes the space in comparison to time.

A retreat, however, shouldn't be made without a good cause such as to save time and force and reallocate the time saved into establishing a new defensive formation or preparing to seize the initiative through counterattack. On the strategic and operational levels, the object of retreating is to "conserve military strength and prepare for the counter-offensive. Retreat is necessary because not to retreat a step before the onset of a strong enemy inevitable means to jeopardize the preservation of one's own forces" (Mao 1963, p. 111). Strategic defense didn't mean surrendering the initiative nor remaining passive, but included offensive operations and re-seizure of the initiative as circumstances and time permitted (Scobell 2011, p. 209). Even though

> we are in a more or less passive position strategically in the first stage of the war, we should have the initiative in every campaign; and of course we should have the initiative throughout the later stages. We are for protracted war and final victory, we are not gamblers who risk everything on a single throw. (Mao 1963, p. 257)

In this Mao echoed Clausewitz who saw strategic defense as a means of protecting national territory and tactical defense as awaiting the enemy attack (Sumida 2007, p. 189). Defensive can employ different means on different levels of warfare. One can use offensive tactics or operational art for strategic defense or vice versa.

Space can be exchanged for time both in offensive and defensive warfare as territory is won or ceded according to the velocity of maneuver and decision-making. Mobility has potential to enhance the strength of defense through utilization of space and trading terrain to time. Simultaneously increased mobility of the attacker erodes the strength of defense. Mechanized war became a competition about which side can best use mobility to its advantage. This was not initially clear to commanders at the dawn of indust-reality. While WW I degraded to passionately holding on to territory seized from the enemy, mechanized warfare of WW II led to a situation where territory gained could just as well be given away again since progress of war was no longer measured by location of frontlines. On one hand Hindenburg (1920, p. 122) had claimed that one doesn't just point at a map and state that he would give up this area. In pre-mechanized warfare one had to think not only as a soldier, but also economically. On the other hand, Hindenburg's experiences in Verdun made him understand that occasionally giving up territory was the most economical solution (von Hindenburg 1920, p. 211). In maneuver warfare

> retreat offers a commander, even with numerically inferior forces, considerable tactical opportunities, […] The farther the enemy advances, and the longer his supply route becomes, the more troops he must leave behind, if he is to be able to maintain himself. During an advance, the supply route is lengthened; during a retreat it is shortened. The retreating army always has its strength concentrated. Hence the moment must eventually come when the retreating is locally superior to its enemy. If at that moment it has access to an adequate supply of petrol and ammunition, it has a wonderful opportunity. It can turn and strike at the advancing enemy and give battle. Such

> an operation must be executed at great speed to ensure that the enemy is given no chance to bring up reinforcements. (Rommel 1953, p. 394–395)

Maneuver warfare has an inbuilt element of bargaining. One must trade terrain to time or strength since growth of distance increases time and energy consumed to traverse it. In controlled retreat one's lines of communication are shortened while the enemy's lines stretch and more energy is consumed in relation to the ground covered. More and more time and energy are required to uphold the momentum of the attack. There comes a moment when the superiority of the advancing enemy turns into inferiority. Acting on that moment and turning the retreat into a counter-attack, one can turn the tables. Fuller (1946, p. 149) argued that in mechanized warfare there should be "no linear or even reliable static defence. Two operations alone are practical—namely, in the offensive to advance, and in the defensive to retire, in order to compel the enemy so to exhaust himself in the follow-up that the initiative is regained." Liddell Hart (1946, p. 22–23) expressed the same idea more graphically. He argued that when the front is too wide to concentrate forces properly, the defending side has to rely on its "own capacity to use space to spin out time, in the hope of exhausting the attacker or of drawing him on to a point where he will be ripe for a counterstroke." The expression "use space to spin out time" is descriptive. Space can be given up to gain time but this equation should weighed against the moral and other repercussions of loss of territory.

Svechin (1992, p. 251) noted that "defensive operations ordinarily involve certain territorial losses. (…) Consequently, for a defense to be successful we must have expendable territory and time must operate to our advantage." When forces are mobile, a retreat must be made in long bounds. To gain sufficient amounts of time to prepare a defensive position, a lot of territory is ceded. Prioritizing time above space is a sound principle, but it requires vast swathes of land that can be traded without endangering strategical objectives. A country such as Russia with extensive territory and the capital far from its borders can economize terrain. As Isserson (2013, p. 47) claimed, "the less territory a country has, the fewer possibilities there are to yield it." It didn't take Guderian's troops long to advance through Belgium and northern part of France to the coast of the channel. Thus, "small countries are ill-suited for positional warfare" (Svechin 1992, p. 254).

What worked for Germany in France failed to work in Russia, partially because of Soviet skill in trading territory for time, but also because of Hitler's timidity in allowing panzer-commanders free rein to rapidly penetrate as deep as possible to conquer the vast emptiness of Russian countryside. Hitler pursued two objectives; strategically to concentrate on the Russian industrial areas forgoing Moscow and operationally to adhere to traditional means of encircling the enemy and creating conditions for the annihilation of the Soviet troops piecemeal (Stempel 2012, p. 156). Instead of encirclements, new corps of commanders wished to

> drive deep, as fast as possible, and leave the encircling of the enemy to be completed by the infantry forces that were following up. Guderian urged the importance of keeping the Russians on the run, and allowing them no time to rally. He wanted to drive straight on to Moscow, and was convinced he could get there if no time was wasted. Guderian's plan was a very bold one—and meant big risks in maintaining reinforcements and supplies. But it might have been lesser of the two risks. By making the armoured forces turn in each time, and forge a ring round the enemy forces they had by-passed, and a lot of time was lost. (Blumentritt cited in Liddell Hart 1948, p. 188)

Maximum speed and force have to be used in mechanized warfare and to slow the advance reduces momentum and ultimately the depth of penetration. Once German tide had reached its high-water mark, Russia seized the initiative and proceeded slowly but with unstoppable force. Hitler ordered Manstein to hold the positions on the Eastern Front and forbade him to yield an inch. Attempting to stabilize a front and to recreate a strong defensive position requires time. "A strategic aim of restabilisation and gaining time must logically make the defender pay some price. And his only currency is space" (Simpkin 1985, p. 280). Manstein wished to exchange territory for time to build a defensive line further west, but orders deprived him of operational freedom. He wrote that "I should have much preferred to be able to submit plans for successful offensives instead of for now inevitable withdrawals. But it is a well-known maxim of war that whoever tries to hold on to everything at once, finishes up by holding nothing at all" (von Manstein 1982, p. 410).

The Eastern Front ultimately collapsed, and while perhaps this collapse was imminent, it nevertheless occurred sooner because Manstein had to attempt to stem the Soviet tide without loss of territory. Every man German troops lost in each battle of attrition fought with insufficient forces had an accumulating effect. Hitler's stubborn policy to hold the ground at any cost, "resulted in devastating German defeats at the operational and, ultimately, strategic levels" (Vego 2009, p. III–59). The mobility of mechanized forces requires boldness in retreat as well as in advance. Occasionally the lack of boldness costs dearly since time is lost. Manstein used all possible means to curtail the velocity of the Soviet advance. As his opponent, Rokossovsky (1970, p. 252) wrote,

> the enemy, retreating under our armies' powerful blows, destroyed bridges, railways and roads, and it took time to restore them. Our supply lines were stretched out over hundreds of kilometres and could not provide all that was needed to maintain our successful advance. It was time to give our troops an opportunity to regroup and prepare for the coming decisive operations.

Rokossovsky was not a master of maneuver warfare, but excelled in defensive battle before making a counterattack at a suitable time (Sokolov 2015, p. 10). For a short period of time, with heavy losses because of having to fight every step of the way, Germans were able to achieve what one long strategic retreat would have given: the over-stretching of Russian lines of communication. When the Russian advance faltered, Germans could have attacked, had their forces not been severely decimated by then. To make a comparison with history and Napoleon's Russian campaign:

> in that pure land war the Russian delaying and waiting tactics were completely successful. Napoleon's army was overtaken by the winter for which it was not prepared and the severity of which it did not imagine, despite many warnings. The nature of the country, its climate, and its inhabitants combined and completed the defeat of the enemy. (Guderian 1956b, p. 128)

For both Hitler and Napoleon, the severity of Russian winter inflicted a mortal surprise, but the unsurpassable enemy was the vast emptiness of Russia. Soviets used the terrain to their advantage. They delayed, waited, retreated, and fought along the way and the momentum of the invader was exhausted. In both cases the drive continued deep into the heartland, all the way to the gates of Moscow, but by that time the energy was spent, the tide crested and started to withdraw. In the end, space conquered mobility since forces simply stretched too thin and lines of supply and communication became too elongated to function. Nevertheless, extensive land area is not a guarantee of defensive success of any state, since the armed forces require skill to utilize existing conditions in its operational art and the state must be strong "to survive the material losses associated with an enemy offensive and make time work in its favor rather than the enemy's." Tactics of trading territory require ruthlessness toward the civilian population of the area given to the enemy. But despite the devastation of the ceded areas, the end justified the means (Svechin 1992, p. 251). Time is more important than space since spatial distances can be overcome with speed, but extensive space requires sufficient mass coupled with mobility to concentrate force where and when necessary.

2.6 Force versus. Mass—From Quantity to Quality and Back

In the writings of many Western theorists of mechanized warfare the importance of quality of troops surpasses their quantity. Nevertheless, this is an illusion. Mass armies create and suffer massive casualties at least in part because of the low level of professionalism among mass of conscripted and quickly trained soldiers. It is not a sound policy to discard demands for neither quality nor quantity. As Svechin (1992, p. 179) wrote, "It is hardly true that there is such a thing as too many troops or troops who are too good. There is no such thing as too many crack troops." The old debate concerning the benefits of professional armies versus mass armies of reservists often uses the example of small Greek citizen armies against the million-man armies of Xerxes. This is something Delbrück (1990a, p. 69–80) proved to be a myth. Probably there were more Greek soldiers than Persian ones, but the victory of the Greeks is a praiseworthy military feat even when stripped of its mythical elements. Naturally the level of training and professionalism has a large role in the result of any battle, but the idea of thousands emerging victorious from a confrontation with millions is self-deception likely to lead to a military disaster. From this military myth springs the idea that the criteria of quality and quantity are incompatible. But masses can be of very high quality as well.

In the agrarian age the quality of warriors had a priority over their number. As Sun Pin (1995, p. 92) wrote, "*enlightened rulers and knowledgeable generals do not rely on masses of troops to seek success.*" Skill was considered to be the most important characteristic of a soldier. This idea prevailed throughout the centuries and perhaps peaked in the era of chivalric warfare when the quality of the limited number of knights was far more important than the mass of peasants opposing them (Delbrück 1990c, p. 283). Similarly the eighteenth century focused on marching and maneuver as the high acme of art of war and everyone preferred quality over quantity to ensure mobility of their troops (Svechin 1992, p. 180). While large armies are difficult to move and to feed, smaller armies are more easily defeated and especially in the Middle Ages small armies were often a necessity dictated by economic reasons (Delbrück 1990c, p. 283).

But even back then the incomparability of numbers and skill was recognized. As Bülow (2013, p. 189) wrote, "victory is decided by number, and not by courage and skill in Tactics. But that number must be conducted with ability; for, when the fronts of armies meet in battle, the more disciplined will, no doubt, put the less to flight." Mass armies of the modern era were created by turning citizens into soldiers through conscription and the resulting casualties led Liddell Hart (1950, p. 338) to argue that "it fosters the fetish of numbers" and that "conscription has been the cancer of civilisation." Growth of both masses and their mobility has had consequences on the development of the art of war and on the impact wars have on the societies waging them.

Increased mobility in warfare has never been only a positive thing, since it enabled war to engulf and exploit larger areas of territory. Mobile troops have always consumed more resources and caused wider destruction. When wars were fought in confined areas between professional armies the amount of devastation was spatially limited. Allegedly nothing grew where the hooves of Attila the Hun had touched the ground. This is because he and other nomadic warriors used the land to feed and support them. This was a pattern specific to not only horse armies but all agrarian invading armies. As de Jomini (2007, p. 103) described it, "Caesar said that war should support war, and he is generally believed to have lived at the expense of the countries he overran." Invasion made the enemy pay the majority of costs and suffer the damage since warfare occurred on his territory. But the better the mobility of armies, the wider the damage spread. Napoleon's system of marches increased the mobility and the range of his troops. Even if economic concerns and costs had been set aside, bringing the supplies his armies required from their homelands would have hindered their movement. They had to be fast of foot and thus marched light. de Jomini (2007, p. 100) saw first-hand the effect of Napoleon's armies and argued that

> if the art of war is enlarged by the adoption of the system of marches, humanity, on the contrary, loses by it; for these rapid incursions and bivouacs of considerable masses, feeding upon the regions they overrun, are not materially different from the devastations of the barbarian hordes between the fourth and thirteenth centuries.

Napoleon's marches and foraging provided his troops with operational mobility and a relative edge in speed, but his methods could not be employed in modern times (Smith 2008, p. 38). At the time of mass and mechanized armies the costs of mobilization extend beyond the physical damage caused in fighting. The million-man armies took their manpower out of production in agriculture and industry while the state still had to provide for them, since the armies were too huge and too geographically concentrated to live on the land (von Bernhardi 1914, p. 24–27). Mechanization spread them over wider areas and minimized force concentration, but increased movement required more resources and especially petroleum.

The successes of the massive but mobile citizen armies of Napoleon led to successive generations of military minds being infatuated with his genius. Napoleonic way of war created a new sense of time-consciousness. He was the first commander to begin issuing time-specific orders (Bucholz 2001, p. 12). This enabled him to concentrate mass into the right place at the right time better than his opponents. Some Napoleon's force concentrations show logistic genius in action. If his followers lacked the genius, they attempted to make good by increasing the mass. Napoleon's innovations gave him an advantage over his opponent on the operational level, but on the tactical level his armies were less effective, since they could not always translate mass of men into mass of firepower (Smith 2008, p. 45).

Germany was among countries intent on learning what made Napoleonic warfare so effective and Scharnhorst led a commission determined to adapt the French methods into German context (Paret 2009, p. 86). Clausewitz was among the people chosen for the commission and his theories ended up having a huge impact especially at the beginning of indust-real and mechanized age. According to Fuller's interpretation, Clausewitzian thought tried "to turn the State into a military machine, and at the very time when steam power was beginning to turn it into an industrial machine. Hence onwards, both armies and factories increasingly ceased to be the servants of the people to become their masters" (Fuller 1946, p. 111). Liddell Hart and Fuller witnessed the horrors of what they considered Clausewitzian destructive battles and turned their backs to this philosopher of warfare. They didn't perceive that the fault was not in Clausewitz, but in those who had only read but not understood his logic. This lack of insight led to the birth the concept of "nation at war".

> From 1866 onwards mass armies take the field. The long-service standing army progressively gives way to the short-service conscript. Quality is ousted by quantity and war becomes the affair of the "average man." [...] the larger armies grow the more dependent do they become on industry to equip, arm and supply them both in peace and war. Industry, the postal and telegraph system etc., are organized for war, for a nation in arms demands a nation of armourers and technicians to sustain and maintain it. (Fuller 1946, p. 117)

States could no longer rely on small professional armies and focused on creating quantity. To produce a mass of soldiers, states relied on conscription, which involved all families in warfare. Similarly all resources, workforce, and means of production of the society were harnessed to support the mass army. As quantity replaced quality, art of war diminished in importance. The "conflict of masses is a war of accidents in which genius is out of place. Though the general can still plot and plan, and increasingly must do so, he can no longer lead or command because the masses are too vast for his grasp" (Fuller 1946, p. 120). With the war of the mass, operational artists lost their personal touch. When mass met mobility, a new dilemma emerged.

> From 1870 onwards, a new civilization had arisen in Europe, based on the enormous growth of railways and the facilities rendered possible by the motor car and lorry. Soldiers had studied these means, not in order to mechanicalize armies, that is to replace muscular by mechanical power, but from the point of view that these means of movement would enable an enemy's frontier to be submerged under a veritable inundation of flesh. Millions of men would sweep forward and, like immense clouds of locusts, would gain victory by sheer weight of numbers. (Fuller 1923, p. 76–77)

The possibilities of mechanization were not wholly understood. It was of no operational or tactical use to increase the concentration of troops near the battlefield if there was no means of fulfilling the rule Henry Lloyd proposed, namely that the "general, who, by the facility of his motions, or by his artifice, can bring most men into action, at the same time, and at the same point, must, if the troops are equally good, necessarily prevail and therefore, all evolutions, which do not lend to this object, must be exploded" (Cited in Liddell Hart 1978, p. 103). The ease of having more and more men at the commander's disposal led to a situation in which some were unable to join the combat. Even with modernized means of transport, it became "difficult to handle armies of millions, to keep them supplied, and to prevent them clogging the arteries of movement. Their very mass stultified the dreams of Napoleonic manoeuvres in which their creators had indulged" (Liddell Hart 1936, p. 22). Operational artists began to search for new ways of concentrating force when they realized they had been beaten by the same concept of mass they had enamored themselves with. As Martin van Creveld (1991, p. 114) perceptively noted, the concentration of force can be created either in space or in time. Out of these, the latter is even more difficult to achieve. This requires mobility and skillful maneuvers. Spatial concentration of forces is inconsequential if they don't arrive at the right time, that is, concentrated temporally as well (Foertsch 1939, p. 42).

In the time of Moltke the Elder the growth spur of the mass armies and their lack of mobility greatly hindered the possibility to practice the art of war through maneuver. Envelopment maneuvers or flanking attacks had brought glorious German victories such as Königgrätz, Gravelotte-St. Privat, or Sedan and Moltke embraced the idea of flanking. Schlieffen even believed the flank attack was the gist of the entire history of war (Vego 2009, p. VII–58). According to Bucholz (2001, p. 3) size, space, and time were paramount in strategic thinking and right combinations had to be created. Yet, as the number of soldiers in an army increased they occupied more space on the battlefield and flanking a mass army required time. Moltke (1993, p. 56) noted that "armies of a hundred thousand and more occupy a space of more than four miles. An envelopment of their flanks becomes a day's march, delays the decision of arms to the subsequent day, gives the opponent an opportunity to evade it." When the battle formation is wide enough, the time to circumvent it becomes long enough for the defender to prepare for the attack.

The mass kept growing toward WW I. As Bernhardi (1914, p. 39) argued, "numbers will therefore always form one of the most essential factors in strategical calculations, and of success. Yet numerical superiority is not always the most important condition for success." What had worked for Moltke, certainly didn't work in the rifle-wars. The idea of amassing troops to the utmost of transport capacity was problematic since when "one man with a machine-gun might count for more than a score, or a hundred, or sometimes even a thousand, who were advancing upon him with the bayonet (…) the formula of victory became merely a formula of futility—and death. The more ranks of attackers, the more swathes of dead" (Liddell Hart 1936, p. 68). Something new needed to be invented. The primary task was to regain protection from fire to plow through the war without exhausting the supply of men a nation could produce for the military machine to grind. Already in 1923 Fuller perceived a lesson from the WW I that proponents of effects-based operations invented anew. He decried the logic behind the mass army and argued that while superiority of numbers traditionally meant men, each man was merely a mounting for a weapon. Thus, the number and potency of weapons was the decisive factor and not the men.

> Men, in themselves, are an encumbrance on the battlefield, and the fewer men we employ, without detracting from sufficiency of weapon-power, the greater will be our concentration of strength, for the aim of concentration is as much concerned with securing an army against blows as it is with enabling an army to deliver them. (Fuller 1923, p. 34–35)

The objective of operational art is to produce as great a concentration of "weapon-power" as possible with as few men as possible to wield the weapons. Writing in the 1920s Fuller propagated in more modern terms concentrating fires or effects and not forces. The fewer men there were on the battlefield, the greater their dispersion and smaller the devastation the enemy could inflict on them. However, to maximize own effects it would be still useful to put into battle as many men equipped with efficient weapons as possibly can partake in it. While the initial idea is sound, the logic of war causes the potential for destruction to grow when more effective weapons are used.

After the fiascos of the WW I the military thinkers began to develop a meaningful new dogma for tactics and operational art. In German circles "complicated theories obstructed the capability to see things in reality. A military dogma had been worked out in every last detail, and this was taken to represent the very peak of military wisdom" (Rommel 2003, p. 96). The new tactics of mechanized warfare was essentially an attempt to make machines perform the task of men, to engage the enemy and pierce his defenses. Superiority in manpower by a horde army could not create a success and German military thinkers from Schlieffen onward recognized that "especially in the period of mass-armies, strategy of annihilation is only possible by continuous movement. Only by movement can rigid fronts be avoided" (Ehrfurth 1991, p. 510). High-mobility maneuver could disturb the stability of trenched fronts. From the perspective of operational art this was an important development, but the German over-emphasis on operational art and sidelining strategy had dire consequences as the first operations had run their course (Gray 2007, p. 107).

Since mass armies led to mass destruction of individual soldiers and mass consumption of resources of societies almost every military theorist in the West—most notably Fuller, Liddell Hart, and De Gaulle—wanted to return to the age of professional high quality armies. They were frightened by the type of army early indust-reality had created. In WW I

> we saw millions and millions of men; such was its main characteristic. We saw the total impossibility of leadership exerting control over these masses. We saw the insuperable difficulties of supplying them. We saw their enormous vulnerability to fire power, and we saw that, like swarms of locusts, they not only destroyed the enemy's country, but devoured the resources of their own. (Fuller 1932, p. 211–212)

Fuller wished to see armies that would remain controllable and could be supplied without an unbearable strain on the economy. Just like Fuller, in Germany Seeckt sought for new solutions. For him the creation of the nation in arms, the mass army, didn't translate to military success. No matter how profound the struggles, the war would not end in the devastating attack or annihilation on the battlefield (Echevarria 2000, p. 226). It would solidify into trench warfare where ultimately one side is no longer able to withstand the casualties and its manpower, material, and morale are extinguished. Seeckt laid down the principles for operational use of mobile units, whether cavalry, motorized, or mechanized (Vego 2009, p. I–22). His professional army was to be well led, well equipped, and able to be maneuvered more effectively than the mass armies (Corum 1992, p. 31).

Seeckt wrote that, "the whole future of warfare appears to me to lie in the employment of mobile armies, relatively small but of high quality, and rendered distinctly more effective by the addition of aircraft, and in the simultaneous mobilization of the whole forces" (Cited in Liddell Hart 1948; See von Seeckt 1930, p. 69). While he emphasized mobility, quality, and co-operation, it was partially due to the fact that building mass was not a viable option. Seeckt's doctrine could be described as "smashing offensive performed with the help of reduced but picked armies—it was imposed by circumstances" (Sikorski 1943, p. 31). Since Germany was unfortified, the prospect of positional warfare was not feasible and planners had to take into consideration the idea of mobile defense with a strong element of offense. Only motorized troops could play a part in a war of movement and Germany embarked on a journey of developing them.

Raising the entire nation in arms, according to Seeckt, had reached the end of its line. The *fureur de nombre* had run its course, mass had become immobile. It could no longer maneuver and reach victory. It was only able to suffocate (von Seeckt 1930, p. 61–62). Suffocating the enemy becomes too costly in terms of men, money, and material, and the principle of the economy of time is violated. Seeckt's operational art depended on having a clear idea of the interrelations of time, force, and space and how to think ahead, simplify the process, and act swiftly to surprise and envelop the enemy (Vego 2009, p. I–22).

Seeckt was forced to favor well-trained and well-equipped small troops for different reasons than the British and French proponents of mechanized warfare. His concept of a decisive and determined offensive with small but select and well-trained units was born out of necessity—there just was not sufficient numerical strength in the remnants of the decapitated German army. He proposed a professional army of twenty-four divisions with 200,000 men (Corum 1992, p. 29). Seeckt's work was a "starting point for a systematic development with a view to the arming of the nation. The German military art has always been based, in fact, on the use of human masses as numerous as possible, as in 1870 and 1914" (Sikorski 1943, p. 31). In this sense, in the interwar period the Germans followed a doctrine of mechanized mass much in the vein of the Soviet thinkers, but didn't have the resources to build the kind of mass they would have preferred. In another sense, they followed the early British views on evolution of a mechanized striking force. They built a great number of simple and relatively cheap tanks instead of striving for highest available technology (Martel 1945, p. 126). As much mass as possible as cheaply as possible and especially as quickly as possible was the paradigm.

The Soviets generally ridiculed the theories of Western thinkers (Simpkin 1987, p. 125, 133). Isserson (2013, p. 9) blamed the bourgeois military thinkers for attempting to replace a "scientific theory of the conduct of operations with vague fantasies on perspectives for future war."

Triandafillov (1994, p. 29) argued that the capitalist countries had become fearful of the proletarian masses and their subsequent revolution. Russian and Soviet strategy and operational art traditionally relied heavily on serfdom and an almost endless resource of human mass (Adamsky 2010, p. 43). Triandafillov acknowledged the objective behind Western calls for reformation, but argued that

> the best conditions for free maneuver, for extensive tactical and operational art, will be achieved not through a return to the small armies of the armchair warriors, but by the corresponding increase in the mobility of modern million-man armies by improving the technology of transportation assets. (Triandafillov 1994, p. 29)

Political viewpoints aside, there is a profound truth in Svechin's (1992, p. 181) statement "in the flame of war small states burn up very quickly." The same applies to its army. A "million-man army" has staying power and a longer duration on the battlefield even if the war proceeds unfavorably. The idea of a great mass was perceived as sound and only the degree of its mechanization was to be increased to add to its combat value.

Another important Soviet theorist to discard western theories was Tukhachevsky, who didn't do justice to the role Fuller's theories played in fueling Soviet thought. He adopted much, but simultaneously argued that the all-arms battle was superior to the tanks-only type of thought abundant in the West (Simpkin 1985, p. 17). He took a different view on the size of the mechanized army than his Western counterparts. Tukhachevsky's vision was of a mass mechanized army with 260 divisions, 40,000 aircraft, and 50,000 tanks which could properly fight out a battle of annihilation (Kipp 2011, p. 71). Tukhachevsky and Triandafillov worked together in developing Soviet operational art and specifically deep operations theory. It was the role of Triandafillov to turn Tukhachevsky's innovative ideas into concrete guidance (Lalu 2014, p. 84, 125).

The West had been abhorred by WW I and attempted to minimize the amount of blood spilled in a future war. Both recognized the need to restore the glory of maneuver to its former place of *primus inter pares* among aspects of operational art. Soviet Union just chose a different path. It wanted to maintain the massive million-man armies, but enhance their mobility. By the end of the 1930s it became clear that the likely enemy would be Nazi Germany and by then its strength was evident to the Soviet leadership. The future enemy was understood to be motorized, armored, and supported with a strong air force (Zukov 1969, p. 172). The difference in perspectives of the Western and Soviet military minds is in preference toward quality or quantity of troops. Soviet doctrine focused on phased mobilization and the 'peak of the war' occurring only after several months of full mobilization (Svechin 1992, p. 200). Toward the end of WW II the victorious armies relied not on quality of small troops operating with boldness, but the superiority of mass, advancing with all bases secured, and avoiding risks. Pre-war theorists in Britain spoke on behalf of highly professional, small armies, but Montgomery with his slow and ponderous maneuvers turned out to be the one commander not once defeated in a major battle during the entire war. De Gaulle (1976, p. 78–79) had envisioned a different type of war:

> One sees, then, how the professional army, ready to march anywhere at any moment, capable, thanks to the internal combustion engine, of reaching the battlefield in a few hours, able to produce every effect of surprise or destruction that it can furnish from the material at its disposal, in short, constructed in all its component parts with a view to obtaining the most complete and the swiftest local results, is in accordance with modern political conditions. There is a grim relationship between the properties of speed, power and concentration which modern weapons confer upon a well-trained military elite, and the tendency of nations to limit the objects of dispute in order to be able to seize them as rapidly as possible and at the least possible cost.

Triandafillov's answer was conservative and harsh. He argued that such a small professional army would risk isolation after penetrating deep into enemy territory unless it was followed by a more traditional massed infantry army for protection. In his argument "a 'ghastly' army also has the resources to battle both tanks and vehicle-mounted infantry. States with million-man armies have all capabilities required not only to drive out, but to isolate and destroy small motorized units that have invaded their territory" (Triandafillov 1994, p. 27–28).

Another point where he didn't see eye to eye with the Western thinkers was the role of the tank in future warfare. While almost everyone in the industrialized West viewed the tank as the future master of the battlefield, Soviets reserved this role for another arm. "Thus, infantry and artillery will comprise the basic mass of future mobilized armies. [...] Infantry and artillery will mainly conduct a future war. Tanks (augmenting and partially replacing artillery) will act in direct concert" (Triandafillov 1994, p. 63). This emphasis on the role of infantry and giving an augmenting role to the tanks results from composition of the Russian society and its culture. Russia was an agricultural society where peasants were an inexhaustible resource. Development of an indust-real mass army out of agricultural people with mass production of materiel started in the 1920s and is occasionally referred to as the first Military-Technical Revolution in Russia. The second such revolution occurred in the 1950s and focused on nuclear weapons and missiles. The third one took place on a philosophical level with the advent of the first clearly Third Wave technologies in the 1970s (Adamsky 2010, p. 26–28). But even mechanization of the army was a demanding task to begin with for what was still mostly a First Wave society. A mass of infantrymen was easier to create in Russia than a mass of machines.

Thus, Triandafillov and other Soviet military theorists took a somewhat Solomonian approach. The debate over the preference of mechanized and sophisticated troops over mass armies was settled by deciding to attempt to have the cake and eat it, that is, by building mechanized mass armies. This decision shaped not only the outcome of WW II but development or armies for most of the Cold War period and proved the truthfulness of Bloch's (1914, p. 58) claim that "mass attracts mass, such is the law of gravitation in war." As long as one belligerent relies on mass, it is almost inescapable that the other side is obliged to resort to it as well. In many cases quality of troops is more important than their quantity and mass, since high-technology weapons and professional soldiers act as force multipliers. But in Russian military thought one of the ways of saving time is to be vastly superior in numbers. Triandafillov attempted to use his methodical calculations in estimating how long a battle would last. He argued that in a battle under

> most favorable conditions (sufficient superiority in forces overall and in suppressive assets in particular, their uninterrupted commitment, highly-trained forces), an outcome cannot be achieved in less than four-five days. If the forces are poorly trained and are not distinguished by special mobility and agility, these time frames can be increased by a factor of 1.5 to 2, even given a sufficient superiority in forces. It goes without saying that outcome peri-

> ods can be extended even more, given any shortage of forces, artillery especially. Given an overall shortage of forces, there may not be any outcome: the operation will founder. (Triandafillov 1994, p. 108)

The logic is that one must be superior to his enemy in numbers, in firepower, in mobility, and in training. The best way to save time in battle is to strive to be superior in every sense, but as Ehrfurth (1991, p. 394) wrote, "absolute superiority everywhere is unattainable; hence it must frequently be replaced by relative superiority somewhere. To achieve relative superiority somewhere is the main objective of almost all military movements and the essential purpose of generalship." To achieve necessary conditions for breaking through the enemy defense, such spatio-temporal superiority is created by concentrating one's forces or their fires. Furthermore, rapid breakthrough in the initial stages of a battle is important to win time. Immediately when combat commences,

> the defender adopts a whole series of measures to reinforce the forces under attack. As indicated above, immediate reserves initially will flow there. Deeper reserves and even forces taken from other fronts (or sectors of the front) will also flow there if the site of the unfolding events is in a sector important from the standpoint of the conduct of the war (or of operations). Duration of an operation greatly depends on the number and rate of accumulation of new forces by the defense. If reserves begin to arrive immediately and in sufficient quantity, an operation may then enter a new phase and last a very long time. (Triandafillov 1994, p. 110)

The Soviet theorists added momentum to the mass. They employed motorized mass armies in deep operations so that the sheer weight and number of advancing units would increase the pressure on the defender and prevent him from bringing the offensive to a standstill. The mass of a freight train, no matter how one attempts to halt it, keeps it grinding forward for a considerable distance. Relying only on quality would be harmful in the long run since the combat value of both armies levels out as professionals are diluted with replacements and the initially inferior mass becomes battle hardened (Svechin 1992, p. 181). Early on in the war the professional army can

> pass in a single bound from peace to war, capture valuable spoils and spread confusion among the enemy during his mobilization period. Their objec-

> tives will, of course, be limited by the means which will have to achieve them. It will not be a question of destroying, by this initial attack, all the forces of the enemy, but one of getting in the first blow. In modern conflicts, where everything has its percussions and repercussions, it is well to show one's determination and to spread anguish beyond the frontier at the first opportunity. (De Gaulle 1976, p. 133)

To summarize, the problem of quality versus quantity is intimately related to time. Time is lost in mobilizing and supplying the huge reservist army, and in the beginning, it's ill-suited for rapid operations. The small professional army loses no time in the early stages of a war. There is no extensive mobilization, no time lost in training, and the army is combat-ready. If a nation has the resources and the will, the time-winning solution would be to model its army along the lines proposed by Triandafillov, but since this rarely is an option, one must include small, mechanized, and professional elements in the armed forces to initiate action and reservist component to be mobilized under the protection of the professional part. The ratio of these two forces should be carefully calculated so that no time is lost in getting the field army operational but bearing in mind the relative costs of both components. The best compromise is a marriage of rapid operability with optimal cost-efficiency. Time is money and in this case to minimize lost time at the beginning of a war would mean spending an exorbitant amount of money.

The efficiency and destructive potential of our contemporary weapons coupled with increased respect for human lives ensure that the era of the 'million-man armies' has come to an end. However, even at the dawn of the post-Cold War era, Odom (1993, p. 60) argued that "quantitative factors will also continue to define the nature of modern warfare. High quality but small forces stand no chance against larger high-quality forces." Quantity is still meaningful, but quality of the troops is crucial. Vego (2009, p. V–56) estimates that "in the future, major land operations will most likely be conducted with much smaller but more capable forces than they are today." Future forces will be equipped with high-technology weapons, they will be better trained and more professional, but there still will remain a considerable number of men fighting our wars. Again, it is not about quality versus quantity but finding a balance between the two and their respective benefits. Perhaps the bottom line still is that "reality provides a very firm answer to the question: one should not sacrifice quality or quantity too much" (Svechin 1992, p. 179).

2.7 Force and Time—Simultaneous or Successive Use

Economy of force as a principle leads to spatio-temporal compression of violence. The quicker, faster, and stronger the attack is, the more violent the shock effect and the more likely a quick result becomes. Compression of force causes a steep but temporary rise in casualty rates. A rapid attack generally entails fewer losses on the whole, although the latter may appear appalling for the time being. Rapidity is an element of particular importance in the tactical offensive. Frederick the Great thus teaches in his General Principles of War: "Therefore, the sharper the attacks are, the fewer men they will cost" (Goltz 1906, p. 157–158).

Economy of force thus resonates in economy of casualties on both sides if the principle is applied correctly. WW I is full of examples where the principle was misunderstood. In Flanders, where the British defense was strongest, Ludendorff wished to gain victory "by closely connected partial blows so that sooner or later the whole structure would collapse... It will be an immense struggle that will begin at one point, continue at another, and take a long time; it is difficult, but it will be successful" (Cited in Goodspeed 1966, p. 241–243). As a result of uneconomical use of force Ludendorff eroded the British defense but simultaneously bled his own troops white at a time when Prussia could ill afford it. While the principle of economic use of force dictates that there must be enough force to ensure the desired effect, it is useless to amass so much force that congestion prevents some elements from participation in combat. This suggests that there is a specific amount of force to be employed economically that varies in each situation and

> beyond this, increase in mass contributes nothing in theory, and brings a nest of problems in practice. One thus arrives at a concept of a 'sufficient mass' which it is normally pointless to exceed at all, and always pointless to exceed by more than a modest margin of insurance. (Simpkin 1985, p. 134–135)

de Jomini recognized the importance of avoiding exaggerations in concentration. According to him one should "keep the mass of the force well in hand and ready to act at the proper moment, being careful, however, to avoid the danger of accumulating troops in too large bodies" (de Jomini 2007, p. 151). Too many troops is harmful since excessive mass slows operations. Maneuverability diminishes making the force unable to answer the demands of the situation. Of whatever mass is available, all that can be utilized in combat is to be used but, as Clausewitz argued, "nothing is more important in strategy than ensuring that the forces that are to carry out an attack are not used in vain, that is to say, that they are not merely thrust into the air" (Cited in Bassford et al. 2010, p. 65).

But it is important to include that "modest margin of insurance" and this category includes the reserves. Every operation requires reserves and their size is based on estimation of the amount of troops that can be brought in to join the fight. The size of reserve should not be exaggerated either. While the massive 1960s' tank armies of the Soviet Union clearly represented an excess of mass, there has to exist a minimum amount of mass as well. The other extreme still sufficient to respond to situational changes was a

> mobile force with enough mass to ensure flexibility. This sets a lower limit to mass. Thus the need to maintain tempo and concentration in time on the one hand, and the need for flexibility on the other, impose upper and lower limits of mass for a given mission and situation. (Simpkin 1985, p. 135)

It is possible to employ too much or too little mass. The principle of economic use is sound but must take into account situational demands that change not only from one time period to another, but according to the type of forces used and, above all, the situation. Achieving just the right amount of mass to be used on any given occasion is an art in itself. Armed forces still recognize the validity of the principle of economy of force. Nevertheless, especially in the smaller states this principle often is interpreted falsely. According to the original idea, economy of force

> is the art of pouring out all one's resources at a given moment on one spot; of making use there of all troops, and, to make such a thing possible, of making those troops permanently communicate with each other, instead of dividing them and attaching to each fraction some fixed and invariable function; its second part, a result having been attained, is the art of again so disposing the troops as to converge on, and act against, a new single object. (Foch 1920, p. 51)

The original idea of economy of force was to concentrate as much force as possible at the same time in the same place. Today, it is often interpreted as using the smallest amount of force possible estimated to still ensure victory (Heuser 2002, p. 86–88). The less force one uses, the smaller is the shock impact on the enemy and when the odds are not overwhelming at the decisive point, it will prolong the battle. And as the duration of battle increases, so do the casualties. Economy of force meant using a vast numerical superiority to gain a quick victory. This example is only one misinterpretation of the principles of war and their utilization in a specific time and place. When Frederick's idea of the oblique order in employing mass was later preached as the infallible rule of winning all battles de Jomini (1992, p. 57) commented, "there is but a step from the sublime to the ridiculous." Frederick used two echeloned lines and made the attack efficient through precision and discipline. This became an established practice and ended as a parody of itself with soldiers advancing in diagonal step (Colby 1943, p. 89–90; van Creveld 2008, p. 355–356).

Occasionally classical texts provide strange guidance of how to employ force. An example of this is attacking simultaneously everywhere and still emphasizing force concentration. Foch is among those who often contradict themselves. He wrote about concentrating all force at the decisive point but argued that "the attack on the whole front must, of course, be resumed at the same time. All the troops of the preparation then turn to execution" (Foch 1920, p. 348). If one orders the entire front to attack, force is dispersed and casualties pile up where the offensive doesn't have clear superiority. Undoubtedly the entire defense is engaged, but there is no operational reason for it. Soviet theorists emphasized the idea of consecutive operations. These consisted of multiple waves of attack in close temporal succession. The theory of successive operations in the 1920s laid the foundation for the theories of deep battle and deep operations of the 1930s (Glantz 1991, p. 1213). The idea of deep battle originated in the tactical level and was a first stage in creating a theory of deep operations (Simpkin 1987, p. 37). Soviet operational art invented the concept of deep operations closely following each other and building on the successes of each preceding operation (Lalu 2014, p. 125–146). The novelty in the concept was the idea of conducting several consecutive attacks, since art of war has always emphasized that

> everything has to be subordinated to the intention of bringing "the maximum of force into battle" at the decisive point. It is a law that heavy blows must be concentrated in space and time. All the advantages of surprise are

> sacrificed if one attempts to reach victory not by one big blow, but by several simultaneous and successive actions. (Ehrfurth 1991, p. 517)

If there is excessive force that can't be engaged in combat simultaneously, it can be used in succession when the operation is carefully timed. Vego (2009, p. V–25) credits Fuller with the first theory of deep operations but also Goltz had already written about the same idea. "The rapid repetition of blows considerably adds to their weight, since each blow is not merely felt for the moment, but its after-effects last a considerable time. The avalanche increases in size and weight only whilst it is rolling, and this applies to success in war" (Goltz, 1906, p. 218). The Soviet thinkers solidified an established idea into doctrine. There rarely are truly novel innovations in warfare, but more or less adequate interpretations, reformulations, and adaptations of old principles. The maxim of spatio-temporally concentrated heavy blows has essentially remained unchanged.

"Huge, multimillion-man armies, fully equipped with modern armaments, have no other prospects for use on the contemporary field of titanic battle, except those delineated by the concepts of the deep operation" (Isserson 2013, p. 3). Mass army had potential for a huge reserve, and mobility coupled with the principle of concentration made it possible to create a second spearhead so that the attack consisted of two echelons. This was meant to exert additional pressure at the same point if the first force concentration had failed. If it had succeeded in creating a breakthrough, the second echelon should continue the attack into the depth. New troops can be used consecutively to cause cumulative damage. Nevertheless, to maximize the edge gained, the enemy must not be allowed any period of recuperation from the initial attack. The consecutive follow-up attack must be in immediate succession. Timing the attacks is the duty of the operational artist since he needs to identify the most auspicious moment in the ebb and flow of the battle.

> The unbroken continuance of the operations demands great intellectual energy on the part of the general. We must remember, that whilst the war lasts, there is for him literally not an hour of rest, not one in which the responsibility weighing upon his shoulders slumbers. (Goltz 1906, p. 218–219)

Guderian was a practical soldier accustomed to presenting his ideas in short and clear form. As he lacked overwhelming mass, Guderian had to maximize the impact of force in one point in time and space. When it comes to the prerequisites for successful offensive he wrote that "we may summarize the requirements for a decisive tank attack by the concepts of: *suitable terrain, surprise and mass attack in the necessary breadth and depth*" (Guderian 1992, p. 181). Surprise is connected to speed of the attack and directly stealing time from the enemy. Breadth and depth are fundamental dimensions in land warfare. Mass acts as a force multiplier and, as Guderian perceived, is diminished by using too much breadth in the attack. There is no rule of thumb and the operational artist must balance the factors, so that mass and speed suffice for a deep localized penetration and not strive for a wide breakthrough. Widening the area of attack just as elongation of the duration of the attack disperses mass and deducts from the momentum needed to penetrate deep.

A powerful crushing blow normally requires, first, that the attacker's pace does not lag behind the rate of withdrawal of the defender's main forces, and second, that the attacker has been provided with the capability to penetrate into the depth of the enemy territory to a distance equal to the length of the enemy front under attack. Operations designed against an enemy occupying a front of 350–400 kilometers require a depth of at least these 350–400 kilometers of accelerated pace (Triandafillov 1994, p. 148). This gives us a scope of the depth one had to be able to penetrate to really inflict serious operational damage in the depth. Thus, the initial breakthrough must be deepened and extended by successive operations since the depth Triandafillov required could be reached only by several consecutive attacks. Immediately when the momentum of the previous attack is spent, fresh units must continue the advance toward the ultimate objective. This type of offensive wouldn't have been applicable prior to mechanization due to insufficient mobility. Muscle power couldn't bring the second attack to bear on the enemy quickly enough nor could it propel the troops into operational depth. The rationale of deep operations was the employment of successive forces to maintain the blow the defender suffers constant for a prolonged period of time. A mechanized

> operation will no longer be a broken chain of interrupted battles. It will be a continuous chain of merged combat efforts throughout the entire depths. It will be a vast sea of fire and combat, spreading across the front as in the World War, but blazing through the entire depths in future war. (Isserson 2013, p. 47–48)

With mechanized troops, maneuver was rapid enough to ensure that the first and second shock wave of attack followed each other practically without delay. If the waves were properly timed, the initial shock brought on the defender would not have time to abate. Rather its momentary nature would be upheld. This shouldn't be interpreted as exerting constant pressure on the enemy but as prolonging the impact of the shock. There is a difference. The edge gained is immediately lost if temporal succession of attacks is not immediate. If the enemy has repulsed the first wave of attack and is granted time to recover from its shock, a second attempt doesn't hold greater chances of success. However, by prolonging the shock, making the 'instant last longer', the devastating effects of shock and surprise accumulate. A shock effect is valuable in operational art because

> decisive results come sooner from sudden shocks than from long-drawn pressure. Shocks throw the opponent off his balance. Pressure allows him time to adjust himself to it. That military lesson is closely linked with the general experience of history that human beings have an almost infinite power of accommodation, to degradation of living conditions, so long as the process is gradual. (Liddell Hart 1946, p. 25)

The more rigid the operational plan of the enemy is, the greater the impact of shock. Simultaneity or succession multiplies the effects. If enemy casualties rise linearly over time, the commander can adjust his operational plan for the altered conditions. A shock effect, due its unexpected and momentary nature, is more likely to derail any operational plan. Accumulation of effects is an attrition-based approach while in decisive deep operations the objective is to synchronize effects so that they occur simultaneously. "The ability to hit many high-value targets simultaneously gives us the wherewithal to employ a strategy of shock and awe that can bring a situation to a conclusion far more rapidly than an attrition-based approach" (Alberts et al. 2000, p. 107).

The shock should be spatially concentrated as well. It seems that the need for temporally concentrated effects is a universal requirement, but the spatial locus of the effects and their quantity depend upon the functions of the targets. Smaller simultaneous effect on important and interdependent targets creates a greater shock than bigger but more dispersed effects. "Shock and awe are achieved not simply as a function of the number of targets destroyed, but as a result of the destruction or neutralization of significant numbers of critical targets within a short period of time and/ or the successful targeting of the right target at the right time" (Alberts et al. 2000, p. 184). Not all targets have to be hit, but crucial ones have to be taken out simultaneously. Shock and awe constituted the U.S. attempt to build a new system of waging war in the 1990s (Kagan 2006, p. 219–233, 261–265). Conceptually it can be viewed as an updated and modified version of the Soviet deep operations. As Tukhachevsky wrote, "modern means of neutralization, employed on a mass scale, put within reach the possibility of simultaneous attack and destruction of the entire depth of the enemy's tactical defence" (Cited in Simpkin 1987, p. 39). Both methods call for widespread attack, but in the Soviet case it was attainable only through maneuver and not precision strikes. Long-distance weapons extended the reach from tactical to operational and strategic levels. But abiding to the idea of deep operations, the shock-effect has to be maintained or recreated time after time with successive waves. This requires skills in temporization and synchronization coupled with a profound understanding of the nature of the enemy to make the right target choices.

No matter how perceptive Triandafillov and other early theorists of deep operations were, they were too tied to the means of past wars to possess the imagination to set forth a vision of future wars. This is evident in the claim that "modern combat is unhurried and is conducted exclusively with rifles, machine guns, artillery, tanks and armored vehicles" (Triandafillov 1994, p. 60). The "unhurried" character was proven untrue time after time in the course of WW II. Speed became the essence of mechanized operational art. Combat was a race of which side was the quickest to plan and execute, and only hurried and elevated pace enabled substantial gains. The attacker had to not only seize initiative but also throughout the operation maintain a higher pace than the defender. Thus, to summarize the operational art in the mechanized age, the "attack must be carried out with utmost speed, in order to take advantage of the surprise effect, penetrate deep into the hostile front, prevent the hostile reserves from going into action, and develop the tactical gains into strategic gains" (Guderian 1937, p. 37).

The theories of *Blitzkrieg* and mechanized operations influenced greatly Cold War operational art. The whole development pattern of the military transformation theories from 1990s onward attempted to make the shift from massive and heavy armies of the Cold War to lighter forces that could be operationally deployed in different theaters around the world (Sloan 2008, p. 5). The cycle of evolution in warfare had made a full turn. The new jargon was similar to that used by Britain and France after WW I when they had colonies all over the world and wished to limit the size of their armed forces. During the Third Wave quality and deployability are favored as timesaving and victory-winning characteristics of the armed forces. Sooner or later the cycle makes yet another revolution and causes us to favor quantity again.

2.8 Winning Time by Increased Mobility and Accelerated Velocity

> Move with lighting speed either to attack or defend. Talent and nimble feet will give you the upper hand. (Ho Chi Minh 2008, p. 27)

While the Vietnamese strategy of protracted war adhered to a long time frame, in tactics and operations all activities had to be rapid. Speed brought protection and increased the impact of attacks for the guerrillas. However, to accomplish more and do it better than the enemy in a given time frame is one of the basic requirements of all types of operational art. Napoleon excelled in controlling movement. He waged war primarily by marching (Delbrück 1990d, p. 423). Napoleon's system was "to march twenty-five miles a day, to fight, and then to camp in quiet. He told me that he knew no other method of conducting a war than this" (de Jomini 2007, p. 100). His forces had a rapid pace of movement that didn't vary, but sufficed to surprise. For Liddell Hart (1932, p. 73) "the quickened rate of strategic and tactical movement was the root of the French successes and made possible that rapid transference of force and bewildering 'reshuffles' of disposition whereby the French multiplied 'mass by velocity.'"

At the time it was customary for armies to march at 70 paces per minute and Napoleon's men substituted it to 120 paces per minute (Liddell Hart 1932, p. 73). They didn't have the practice to "conform to the orthodox slow step and the quick step became normal for marching and fighting—a sacrifice of solidity and symmetry to speed. The quickened rate of movement enabled the French to outmanoeuvre their opponents on the battlefield" (Liddell Hart 1932, p. 66). Napoleon discarded those parts of repetitive drill training that he didn't consider important and focused on enhancing the mobility of his troops. Moreover, the benefit of Napoleon's mobility was two-fold. He exploited it both "as a means to surprise his enemy, and also to ensure his own security" (Liddell Hart 1932, p. 73).

Napoleon was often outnumbered by his enemies and had to rely on skill and speed in his operations. Napoleon's original mathematics of war allowed him to overcome this hindrance of numerically inferior troops. He argued that the strength of an army, "like the power in mechanics, is estimated by multiplying the mass by the rapidity; a rapid march augments the moral of an army, and increases all the chances of victory" (Napoleon 1987, p. 58). If the mass was inadequate, numerical balance could be altered by increased mobility. Napoleon was an offspring of the French Revolution that "introduced the system of divisions, which broke up the excessive compactness of the old formation, and brought upon the field fractions capable of independent movement on any kind of ground" (de Jomini 1992, p. 278). He restructured his armies to operate with smaller formations. He understood that time saved on strategic and operational levels in the movement of armies and on tactical level in smaller and more agile battle formations enhanced overall performance. In his equation mass multiplied by rapidity equaled capability. Time need to be included in all calculations of force.

A profound understanding of the need for speed can be found in Frederick's maxim that "promptness contributes a great deal to success in marches and even more in battles. That is why our army is drilled in such a fashion that it acts faster than other. From drill come these maneuvers which enable us to form in the twinkling of an eye" (Frederick 1985, p. 395). He used discipline and drill to achieve a modest margin in mobility which enabled him to strike before the enemy was prepared. Mobility in tactics and operational art is always relative. One just needs to be able to act faster than his enemy (Colby 1943, p. 80, 145). There is no pressing need to be constantly as fast as possible, since accelerated speed of maneuver increases the potential to blunder. Frederick wanted rapidity to cover all battlefield activity. If every soldier is honed through drills to automatic responses to commands, no time is wasted in execution. This saved time directly diminished the response time of the enemy. Individual solders' speed of action had a tactical effect. With this Germanic tradition of drills one could change battle formations and maneuver them quickly enough to throw the enemy off his balance and rhythm.

During indust-reality the situation changed and speed of maneuver started to become crucial operationally. Since mass was still considered to be a worthy force multiplier in WW I the development between the World Wars was significant. Liddell Hart (1927, p. vi) claimed that, "the smaller our land forces the more essential is it that we should 'multiply their mass by their velocity,' and be able to concentrate them rapidly at any point on the land frontiers." Velocity enabled creating force concentrations rapidly wherever necessary.

By 1935 the British army had conceived of an attack force consisting of fast and lightly armored tanks for mobile warfare and slower and heavily armored ones to support the infantry (Martel 1945, p. 117). Guderian argued based on studying the history of war, exercises of armored troops in Britain, and his personal experience that tanks could reach their full potential only when the other arms they are depended on could become just as maneuverable in all types of terrain. Tanks should in this orchestra play the leading violin and all others should conform to their actions. According to him, tanks should not be attached to the infantry and used piecemeal. There would be no use in creating infantry divisions and suppling them with tanks, but to establish panzer divisions including all support elements necessary for tank operations (Guderian 1956a, p. 20–21).

Guderian argued that the offensive power provided by mechanization rested in balance of firepower, speed, and armor protection. Yet, the most important of the trinity was speed, because "a curtailment in speed, which may be required to retain contact with the less mobile arms, will afford the opponent more time to bring into play his antitank guns as well as to launch a counterattack with armored forces" (Guderian 1937, p. 15). The importance of speed is closely related to time. Time won was stolen from the enemy, and mobility acted as the enabler of operations.

Liddell Hart divided mobility into three factions. These are "guarding mobility" which meant providing reconnaissance and protection to the main body of troops by situating the cavalry as a protective screen at a distance toward the enemy. The second is "strategic mobility" where a commander is able to transfer part of his strength "from one point to another to effect an unexpected concentration of force at some vital spot." Third comes "hitting-mobility—that used for direct offensive action—which lies in the impetus of attack and demoralizing effect given by speed of onslaught" (Liddell Hart 1927, p. 51). The first one is less important than the other two, because the benefits derived from guarding mobility can be gained by other means. It is about preventing the enemy to inflict a surprise and even a static extended protective formation can produce the desired effect. What Liddell Hart termed strategic mobility is today considered operational and is of elevated importance to creating time-sensitive operational art. The last aspect of mobility, hitting mobility, resides entirely within the realms of tactics and operational art. It ensures the troops can use speed to proceed to contact with the enemy, concentrate force to break-through the enemy front, and continue into the depth before the less mobile enemy could curb their impetus.

> Through the tank was reborn the art of surprise, to which it added the relentlessness of machinery. Through it the art of manoeuvring was restored in detail, since it could deliver either a frontal or a flank attack under fire, move and fire at the same time, and advance in any direction. (De Gaulle 1976, p. 67–68)

Fuller and Liddell Hart invented a suitable neologism to describe the advantage mechanized army holds over the traditional troops of the past. This was "*loco-mobility—their ability to move over every sort of ground and to clear every yard of any locality*"" (Liddell Hart 1927, p. 116). Loco-mobility enabled troops to maneuver with fewer restrictions set by features of terrain. Fuller argued that loco-mobility was the essential characteristic of mobility in mechanized warfare. With this he referred to the ability to move flexibly and "freedom of movement in all directions" (Fuller 1923, p. 51). An infantry column had good mobility since marching it could advance fifteen miles a day, "but its loco-mobility—that is, its power to move at right-angles to its line of advance—is negligible" (Fuller 1926, p. 160). The search for flexibility to change direction of the entire force quickly led to a creation of new geometrical fighting formations (Gat 2001, p. 69–96). Concentration, cross-country movement, and changing direction rapidly are all facets of loco-mobility. This led to replacement of line tactics with area tactics where

> the front of an army will no longer so completely protect its rear services and its line of communications as today. Attacks will take place in areas and not against lines: they may come from any direction; therefore power to surprise is vastly increased, and the moral attack will grow in importance, its aim being to effect disorganization through demoralization rather than through destruction. (Fuller 1943, p. 43)

Loco-mobility also entails the ability to move fast with protection. There is no way a tank could compete with a helicopter in speed of movement. But a tank can be employed over a wide range of different terrains with a high level of protection. Installing armor as a protective measure can slow an automobile down and make a helicopter unable to fly. There is an irreconcilable conflict between "the respective claims of mobility and security. In reconciling these claims, it is wise generally to give preference to mobility—as the offensive and time-winning factor" (Liddell Hart 1950, p. 316). Being safe often means acting cautiously and not attempting to audaciously seize the initiative. Demands are partially contradictory and the best possible compromise must the attained, as the Germans managed to do. They proved with *Blitzkrieg* operations the effectiveness of disorganization and demoralization by attacking from an unexpected direction, with unexpected forces or at an unexpected time. This ambiguousness in predetermining future actions thickened the fog of war. Fuller wrote that,

> success in war depends upon mobility and mobility upon time. Mobility leads to mass, to surprise and to security. Other things being equal, the most mobile side must win: this is a truism in war as in horse-racing. The tank first of all is a time-saving machine, secondly a shield—it is, in fact, an armoured mechanical horse. If in a given time we can do three times as much as the enemy and lose a third less than he does, our possibilities of success are multiplied by nine. (Cited in Carver 1979, p. 27–28)

According to Fuller's logic, time is the key to success, and mobility allows saving time and striving for greater successes. Mobility is a timesaving force multiplier since the more mobile counterpart is able to perform more and waste less time. Mass and mobility are natural enemies. This is why armies never move at the pace their vehicles theoretically could uphold. When highly mobile units join together to create bigger formations impediments to mobility occur. In force concentration mobility causes problems

> as a means of amassing weight more easily—and by multiplying their impedimenta they have subtracted from the addition to their mobility. Even where space is large enough to enable a force to outflank and by-pass centres of resistance, its freedom of movement is limited by its own maintenance needs. (Liddell Hart 1946, p. 19–20)

Open space is no guarantee of the ability to fully employ mobility. The need for supplies and maintenance restricts full utilization of all space available for maneuver. As Fuller (1948, p. 98) argued, "mobility in the attack is superior to mass in the defence." One used all possible methods to slow down or halt the enemy with obstacles, minefields etc. while attempting to speed up one's own maneuvers—and naturally countering the resistances the enemy will on its turn inflict on free and rapid movement. Thus, on the operational level movement is in some parts of the theater reduced and simultaneously increased in others. There co-existed areas in a state of stagnation and rapid movement. In both the enemy attempted to alter the status, turning the stasis into motion or halting the movement to a full stop, if possible. However, "in the secrecy and the speed of strategical movement now, even more than in the past, lied the foremost conditions of success" (Sikorski 1943, p. 151). Tactics of the defender will attempt to win time by reducing the enemy his speed but the attacker needs to strive for speed and consume as little time as possible in his bid for victory. For all belligerents the primary characteristic of indust-real warfare was

> movement. And as to this there can be no doubt. Though production has vastly increased, it is movement, in its many forms of steamship, railway, motor-car, telegraph, telephone and wireless transmission, which more than any other factor has created the modern world. Movement is the vital element in what we call Western civilization, and restriction of movement its death element. (Fuller 1932, p. 84)

Since physical movement characterized the indust-real Western way of life, it was not surprising that maneuver occupied such a central role in operational art as well. To counter the increased mobility in warfare, one has two options: either to be faster and more mobile than the enemy or to reduce the enemy's speed and mobility. The former is problematic, because there are practical limits to speed, since pace of operations needs to remain within bounds of human capability to control it. The latter is not always suitable either, since ultimately the reduction of speed will lead to bloody battles of attrition. Therefore, the answer was to be found in the combination of both. In the agrarian First Wave the great captains were mostly able to save time in tactical movement. Indust-real Second Wave saw the impact of accelerated velocity on operational movement within a theater of war. During the Third Wave operational velocity has still increased, but the effects start to resonate on the strategic level.

If indeed the glory belongs to the active, "a general who moves his masses rapidly and continually, and gives them proper directions, may be confident both of gaining victories and of securing great results therefrom" (de Jomini 1992, p. 176). One must not focus only on "rapid and continuous" movement of the masses, but on the idea of "proper" directions. Increased mobility and maneuver will lead to disorderly or chaotic situations unless strict control is exerted and the "directions" given must be well-coordinated and immaculately synchronized. Furthermore, since the enemy exerts his influence one needs to keep maneuvers simple and quick to execute under pressure. Movement and maneuvers are necessary, but only when used to deliver a powerful stroke (Colby 1943, p. 147).

The art of war consists of the mass of troops purposefully and concentrating their movements to "act at the decisive moment and at the decisive point of the field of battle" (de Jomini 1992, p. 322). Fuller argued on behalf of movement as the prominent factor in the plans of operational artists of the mechanized age. It didn't matter whether one initially took the defensive or the offensive stance, he could make it or brake it if he upheld enough purposeful movement because

> future plans will have to be based more on movement than on offensive action; by which I do not mean that the offensive will become less important, but that its effect will be the more dependent on correct and rapid movement. Once the opportunity to strike offers itself, movement must be immediate and by signal and not by operation order. (Fuller 1943, p. 49)

Controlling and regulating movement is a burden of operational artists since they must plan and execute maneuvers as the situation calls for. Offense or defense, the effect is tied to rapid maneuver, at the right time, into the right place. The immediacy Fuller calls for when the moment is ripe can only be reached through preparation. Once the time to act comes, orders must be already distributed to the troops and just their execution initiated. The demand for mobility extends to the commander who must be in the right place at each moment. "The correct position for every commander is at the point of greatest importance, a point which is constantly changing in its locality" (Fuller 1943, p. 53). In maneuver warfare there no longer exists an Archimedean immovable point on the battlefield for the commander. Everything is in motion and the motion should accelerate constantly to win time. But just as Fuller emphasized the proper directions and purposeful movement, the same applied to speed. "Organized velocity" was supposedly the secret (Fuller 1948, p. 80). This makes synchronization of mobile elements extremely demanding.

> Forces must attain maximum successes, maximum movement forward. The enemy will attempt especially strongly to defend vital axes, vital points, which cover his withdrawal routes, or a concentration of fresh forces. The art of the attacker is to determine these axes and these points and to unleash the entire mass of forces quickly enough to break out to the flank and rear area of the enemy forces, cut his withdrawal routes and disrupt any new grouping of forces the enemy is preparing. Combat actions during this period require an incredibly fast pace, the maximum possible intensity, great flexibility, and maneuverability. (Triandafillov 1994, p. 49)

This quotation illustrates a desired end state of what could theoretically be obtained with mobile forces. It wasn't a depiction of Soviet mechanized warfare at the time of writing, but a hopeful vision of the future. In real life, Clausewitzian fog of war and friction create a situation where attacks will not proceed unhindered. Friction distinguishes real war from war on paper and the operational artist has to account for it (Howard 1983, 26). There will inescapably be halts and pauses, but they can be managed. The pace of movement doesn't need to be steady and persistent. There are occasions, such as the follow-through, when the speed of the mechanized formations should be as high as possible for maximum penetration. During the actual breakthrough attempt constancy of speed is not required. For this, Liddell Hart (1950, p. 288) offers clear advice: "Pace with variability is the secret of mobility, and sustained momentum, in the follow-through." Going full speed ahead does not epitomize mobility, but rather the ability to use varying speeds and sustain momentum of attack. Mobility allows for variations in tempo and adds uncertainty since speed of maneuver becomes unpredictable.

The Soviet concept of deep operations adhered to the idea of pace with variability even if this was not explicitly stated. Until approximately the 1960s the deep operation theory continued to revolve around a holding force over a wide front, the break-in battle, and deep penetration by mechanized units. As Simpkin (1985, p. 43) described the temporizing and the pace set for this type of attack, "the principle of 'slow in, fast out'—deliberate action/tight rein in the break-in, and dash/loose rein in the break-out—is unchanged." The variability of the pace comes from reliance on attrition in the first part of the attack, slowly and deliberately grinding the enemy. The actual time and place for the planned breakthrough was kept unclear from the enemy, but once combat started, the pace was deliberately slow, making sure that there would be no haste and the break-in would succeed. When the front was breached, speed was accelerated to the maximum and a dash into the depth began. It was the ultimate goal of indust-real operational art that one could say:

> There will then be no halt in the advance, or the attack, in a pursuit, or in a retirement; therefore no time will be lost. In brief, to economize time in action will become the soul of every plan. (Fuller 1943, p. 15)

This chapter illustrated how the factors of time, space, and force could be joined together. Despite numerous attempts throughout the ages these quantifiable factors cannot be forced into calculations and equations that would produce an infallible and universal winning formula for war. The human factor complicated the equations by adding too many variables. The equilibrium between time, force, and space is created anew by the operational artist in each situation, and only generalizations can be made as guidelines to support creating a favorable balance.

This chapter has attempted to show that in the history of operational art, the pursuit for the 'perfect' balance of time, space, and force has never ceased. The development pattern has been cyclical since any alteration in one of the factors, such as the increase or decrease of mass in forces has affected the utilization of space and time. As mass grows, mobility is reduced, time is lost, and space is potentially ceded to the enemy. If time is wasted the enemy can increase his force and win more space. But space can be willingly given away in exchange to time. The trinity is interdependent, and creating an overall balance favorable in a given situation is the task of the operational artist. The mastery of one requires a profound understanding of the others.

All in all, the relationships of time, space, and force are complex. Abstract and imaginative thinking instead of dogmatic rules is required to conjoin these physical factors with immaterial ones such as intellect, imagination, and morale, and weave them into operational art. Mastery of the trinity of time, space, and force is a prerequisite of a commander, and his ability to balance them optimally is the pinnacle of operational art. The trinity must be composed differently for every occasion. As an example, the roles of time and space are different for the attacker and the defender. Thus, the next chapter will discuss how they should be used for different purposes. Activity should be controlled within the temporal frame by managing the speed of action and trying to identify the most promising moments to initiate action. It is not enough to have unrivaled speed or absolute superiority in mass, but to control them and employ them to gain an advantage.

References

Adamsky, D. (2010). *The Culture of Military Innovation – The Impact of Cultural Factors on the Revolution in Military Affairs in Russia, the US, and Israel.* Stanford: Stanford Security Press.

Alberts, D. S., Garstka, J. J., & Stein, F. P. (2000). *Network Centric Warfare – Developing and Leveraging Information Superiority.* (2nd ed. (revised)). Washington: CCRP Publication Series.

Bassford, C., Moran, D., & Pedlow, G. W. (2010). On Waterloo – Clausewitz, Wellington, and the Campaign of 1815. Clausewitz.com.

Bazilevich, K. (1945). The Russian Art of War. *Soviet War News.*

Beaufre, A. (1965). *An Introduction to Strategy – With Particular Reference to Problems of Defense, Politics, Economics, and Diplomacy in the Nuclear Age.* Praeger/New York/Washington: Frederick A.

Bellamy, C. (2007). *Absolute War – Soviet Russia in the Second World War.* New York: Alfred A. Knopf.

Bond, B. (1998). *The Pursuit of Victory: From Napoleon to Saddam Hussein.* Oxford: Oxford University Press.

Brodie, B. (1959). *Strategy in the Missile Age.* Princeton.: Princeton University Press.

Brodie, B. (1973). *War and Politics.* London: Cassell.

Bucholz, A. (2001). *Moltke and the German Wars, 1864–1871.* New York: Palgrave Macmillan.

Carver, M, Field Marshal Lord. (1979). *The Apostles of Mobility. The Theory and Practice of Armoured Warfare.* London: Weidenfield and Nicolson.

Colby, E. (1943). *Masters of Mobile Warfare.* Princeton: Princeton University Press.

Corum, J. S. (1992). *The Roots of Blitzkrieg - Hans von Seeckt and German Military Reform.* Lawrence: University Press of Kansas.

de Bloch, J. (1914). *Future of War in Its Technical Economic and Political Relations.* Boston: The World Peace Foundation.

De Gaulle, C. (1976). *The Army of the Future.* Westport: Greenwood Press.

de Jomini, A. H. (1992). *The Art of War.* London: Greenhill Books.

de Jomini, A. H. (2007). *The Art of War.* Rockville: Arc Manor.

Delbrück, H. (1990a). *Warfare in Antiquity - History of the Art of War* (Vol. I). Lincoln/London: University of Nebraska Press.

Delbrück, H. (1990b). *The Barbarian Invasions - History of the Art of War* (Vol. II). Lincoln/London: University of Nebraska Press.

Delbrück, H. (1990c). *Medieval Warfare - History of the Art of War* (Vol. III). Lincoln/London: University of Nebraska Press.

Delbrück, H. (1990d). *The Dawn of Modern Warfare - History of the Art of War* (Vol. IV). Lincoln/London: University of Nebraska Press.
Dillon, M., & Reid, J. (2009). *The Liberal Way of War. Killing to Make Life Live.* London: Routledge.
Douhet, G. (1999). The Command of the Air. In *Book 4: Roots of Strategy* (pp. 263–408). Mechanicsburg: Stackpole books.
du Picq, A. (1987). Battle Studies – Ancient and Modern Battle. In *Book 2: Roots of Strategy* (pp. 65–300). Harrisburg: Stackpole books.
Dupuy, T. N. (1987). *Understanding War – History and Theory of Combat.* New York: Paragon House Publishers.
Echevarria, A. J., II. (2000). *After Clausewitz – German Military Thinkers Before the Great War.* Lawrence: University Press of Kansas.
Echevarria, A. J., II. (2013). *Clausewitz and Contemporary War.* Oxford: Oxford University Press.
Ehrfurth, W. (1991). Surprise. In *Book 3: Roots of Strategy* (pp. 351–554). Mechanicsburg: Stackpole books.
Foch, F. (1920). *The Principles of War.* New York: Henry Holt and Company.
Foertsch, H. (1939). *Nykyinen ja tuleva sotataito* (Org. Kriegskunste Heute und Morgen) Werner Söderström Oy, Porvoo.
Franks, T. (2004). *American Soldier.* New York: Regan Books.
Frederick The Great. (1985). Military Instructions for the Generals. In T. R. Phillips (Ed.), *Roots of Strategy* (pp. 301–400). Harrisburg: Stackpole Books.
Freedman, L. (2006). *The Transformation of Strategic Affairs.* (Adelphi Paper 379). London: The International Institute for Strategic Studies.
Fuller, J. F. C. (1923). *The Reformation of War.* London: Hutchinson.
Fuller, J. F. C. (1926). *The Foundations of the Science of War.* London: Hutchinson.
Fuller, J. F. C. (1932). *The Dragon's Teeth – A Study of War and Peace.* London: Constable.
Fuller, J. F. C. (1943). *Armoured Warfare – An Annotated Edition of Fifteen Lectures on Operations Between Mechanized Forces.* London: Eyre & Spottiswoode.
Fuller, J. F. C. (1946). *Armament and History – A Study of the Influence of Armament on History from the Dawn of Classical Warfare to the Second World War.* London: Eyre & Spottiswoode.
Fuller, J. F. C. (1948). *The Second World War 1939–45. A Strategical and Tactical History.* London: Eyre & Spottiswoode.
Fuller, J. F. C. (1960). *The Generalship of Alexander the Great.* New Brunswick: Rutgers University Press.
Fuller, J. F. C. (1961). *The Conduct of War 1789–1961: A Study of the Impact of the French, Industrial, and Russian Revolutions on War and Its Conduct.* London: Eyre & Spottiswoode.

Gat, A. (2001). *A History of Military Thought from the Enlightenment to the Cold War*. Oxford: Oxford University Press.

Glantz, D. M. (1991). *Soviet Military Operational Art - in Pursuit of Deep Battle*. Oxon: Frank Cass.

Goltz, Field-Marshal Baron von der. (1906). *The Nation in Arms – A Treatise on Modern Military Systems and the Conduct of War*. London: Hodder and Stoughton.

Goodspeed, D. J. (1966). *Ludendorff – Genius of World War I*. Boston: Houghton Mifflin Company.

Gray, C. S. (2007). *Fighting Talk – Forty Maxims on War, Peace, and Strategy*. Westport: Praeger Security International.

Gray, C. S. (2012). *Airpower for Strategic Effect*. Maxwell Air Force Base, Alabama: Air University Press.

Guderian, H. (1937). *Armed Forces and their Cooperation with Other Arms*. Fort Humphreys: The Army War College.

Guderian, H. (1956a). Sotilaan Muistelmia, Org. Erinnerungen Eines Soldaten (Wolf H. Halsti, Trans.). Kustannusosakeyhtiö Otava, Helsinki.

Guderian, H. (1956b). Russian Strategy in the War. In B. H. L. Hart (Ed.), *The Red Army* (pp. 127–133). New York: Harcourt, brace and company.

Guderian, H. (1992). *Achtung – Panzer! The Development of Tank Warfare*. (C. Duffy, Trans.). London: Cassell.

Heuser, B. (2002). *Reading Clausewitz*. New York: Random House.

Howard, M. (1983). *Clausewitz*. Oxford: Oxford University Press.

Isserson, Georgii Samoilovich. (2013). *The Evolution of Operational Art* (B. W. Menning, Trans.). Fort Leavenworth: Combat Studies Institute Press.

Kagan, F. W. (2006). *Finding the Target - the Transformation of American Military Policy*. New York: Encounter Books.

Kaldor, M. (2007). *New & Old Wars* (2nd ed.). Stanford: Stanford University Press.

Kautilya. (n.d.). *Arthashastra*. http://www.columbia.edu/itc/mealac/pritchett/00litlinks/kautilya/index.html. Downloaded 1.6.2020.

Kipp, J. W. (2011). The Tsarist and Soviet Operational Art, 1853–1991. In J. A. Olsen & M. van Creveld (Eds.), *The Evolution of Operational Art – From Napoleon to the Present* (pp. 64–95). Oxford: Oxford University Press.

Lalu, P. (2014). *Syvää vai Pelkästään Tiheää? – Neuvostoliittolaisen ja venäläisen sotataidollisen ajattelun lähtökohdat, kehittyminen, soveltaminen käytäntöön ja nykytilanne. Näkökulmana 1920- ja 1930-luvun syvän taistelun ja operaation opit*. Taktiikan laitos, Julkaisusarja 1 Nro 3/2014, Maanpuolustuskorkeakoulu, Helsinki.

Liddell Hart, B. H. (1927). *The Remaking of Modern Armies*. London: John Murray.

Liddell Hart, B. H. (1932). *When Britain Goes to War: Adaptability and Mobility*. London: Faber and Faber Ltd.

Liddell Hart, B. H. (1936). *The War in Outline 1914-1918*. London: Faber and Faber Limited..
Liddell Hart, B. H. (1946). *The Revolution in Warfare*. London: Faber and Faber.
Liddell Hart, B. H. (1948). *The Other Side of the Hill – Germany's Generals, Their Rise and Fall, with Their Own Account of Military Events 1939–1945*. London: Cassell and Company.
Liddell Hart, B. H. (1950). *Defence of the West – Some Riddles of War and Peace*. London: Cassell and Company.
Liddell Hart, B. H. (1978). *The Sword and the Pen*. Ed. Adrian Liddell Hart. London: Cassell.
Ludendorff, E. (1919). *Sotamuistelmani 1914–1918*. Porvoo: WSOY.
Luttwak, E. N. (1987). *Strategy: The Logic of War and Peace*. Cambridge, MA: Belknap Press of Harvard University Press.
MacArthur, D. (1964). *Reminiscences*. London: Heineman.
Mao Tse-Tung. (1963). *Selected Military Writings*. Peking: Foreign Languages Press.
Martel, G. (1945). *The Problem of Security*. London: Michael Joseph.
Martel, Giffard Le Q. (1931). *In the Wake of the Tank: The First Fifteen Years of Mechanization in the British Army*. London: Sifton Praed.
Matheny, R. R. (2012). *Carrying the War to the Enemy – American Operational Art to 1945*. Norman: Oklahoma University Press.
Minh, H. C. (2008). *Prison Diary*. Hanoi: The Gioi Publishers.
Montgomery. (1961). *The Path to Leadership*. London: Collins.
Napoleon Bonaparte. (1987). *The Military Maxims of Napoleon* (Ed. D. G. Chandler). London: Greenhill Books.
Odom, W. E. (1993). *America's Military Revolution - Strategy and Structure after the Cold War*. Washington: American University Press.
Paret, P. (2009). *The Cognitive Challenge of War – Prussia 1806*. Princeton: Princeton University Press.
Patton, G. S., Jr. (1947). *War as I knew it*. Cambridge: Riverside Press.
Procopius. (1968). *History of the Wars, Books V and VI*. Procopius with an English Translation by H.B. Dewing, In Seven Volumes, Vol III, William London: Heinemann.
Reid, B. H. (1997). *J.F.C. Fuller: Military Thinker*. London: Macmillan Press.
Rokossovsky, K. (1970). *A Soldier's Duty*. Moscow: Progress Publishers.
Rommel, E (1953). *The Rommel Papers* (Ed. B.H. Liddell Hart). Collins: London
Rommel, Erwin. (2003), *Rommel and His Art of War* (Ed. Dr J. Pimlott). London: Greenhill Books.
Rommel, E. (2006). *Infantry Attacks*. London: Greenhill Books.
Schileffen, Count Alfred von. (1936). *Cannae the Command and General Staff School Press*. Fort: Leavenworth.

Schwarzkopf, N. H. (1993). *It Doesn't Take a Hero – The autobiography, written with Peter Petre*. New York: Bantam Books Paperback edition.
Scobell, A. (2011). The Chinese Way of War. In J. A. Olsen & M. van Creveld (Eds.), *The Evolution of Operational Art – From Napoleon to the Present* (pp. 195–221). Oxford: Oxford university press.
Sikorski, General W. (1943). *Modern Warfare – Its Character, Its Problems*. London: Hutchinson.
Simpkin, R. E. (1985). *Race to the Swift – Thoughts on Twenty-First Century Warfare*. London: Brassey's Defence Publishers.
Simpkin, R. E. (1987). *Deep Battle - The Brainchild of Marshal Tukhachevskii*. London: Brassey's Defence Publishers.
Sloan, E. (2008). *Military Transformation and Modern Warfare. A Reference Handbook*. Westport: Praeger Security International.
Smith, R. (2008). *The Utility of Force: The Art of War in the Modern World*. New York: Vintage Books.
Sokolov, B. (2015). *Marshal K. K. Rokossovsky – The Red Army's Gentleman Commander*. Solihull: Helion & Company.
Ssu-ma. (1993). The Methods of the Ssu-ma. In R. D. Sawyer (Ed.), *The Seven Military Classics of Ancient China*. (pp. 107–144). Boulder: Westview press.
Stempel, J. (2012). *The Nature of War – Origins and Evolution of Violent Conflict*. Jefferson: McFarland.
Strachan, H., & Herberg-Rothe, A. (2007). In H. Strachan & A. Herberg-Rothe (Eds.), *Clausewitz in the Twenty-First Century* (pp. 1–13). Oxford: Oxford University Press.
Sumida, J. (2007). On Defence as the Stronger Form of War. In H. Strachan & A. Herberg-Rothe (Eds.), *Clausewitz in the Twenty-First Century* (pp. 163–181). Oxford: Oxford University Press.
Sun Pin. (1995). *Military Methods*. Translated with introduction and commentary by Ralph Saywer. Boulder: Westview Press.
Sun-tzu. (1993). Art of War. In R. D. Sawyer (Ed.), *The Seven Military Classics of Ancient China* (pp. 145–186). Boulder: Westview Press.
Svechin, A. A. (1992). *Strategy*. Boulder: Eastview Press.
T'ai Kung. (1993). Six Secret Teachings. In R. D. Sawyer (Ed.), *The Seven Military Classics of Ancient China* (pp. 40–106). Boulder: Westview Press.
Toffler, A., & Toffler, H. (1995). *War and Anti-war: Survival at the Dawn of the 21st century*. New York: Warner Books.
Triandafillov, V. K. (1994). *The Nature of the Operations of Modern Armies* (W. A. Burhans, Trans). Portland: Frank Cass.
van Creveld, M. (1991). *On Future War*. London: Brassey's.
van Creveld, M. (2005). *The Art of War - War and Military Thought*. New York: Harper Collins Publishers.
van Creveld, M. (2008). *The Culture of War*. New York: Ballantine Books.

Vego, M. (2009) *Joint Operational Warfare: Theory and Practice* (Revised Edition). Newport: U.S. Naval War College.
von Bernhardi, F. (1914). *How Germany Makes War*. London: Hodder & Stoughton.
von Bülow, D. H. (2013). *The Spirit of the Modern System of War*. Cambridge: Cambridge University Press.
von Clausewitz, C. (1989). *On War*. Edited and translated by Michael Howard and Peter Paret (First paperback ed.). Princeton: Princeton University Press.
von Freytag-Loringhoven, H. (1991). The Power of Personality in War. In *Book 3: Roots of Strategy* (pp. 173–349). Mechanicsburg: Stackpole books.
von Hindenburg, M. (1920). *Elämäni*. Porvoo: Werner Söderström Osakeyhtiö.
von Manstein, E. (1982). *Lost Victories*. Novato, California: Presido Press.
von Moltke, H. (1993). *On the Art of War – Selected Writings* (Ed. D. J. Hughes). New York: Ballantine Books.
von Seeckt, H. (1930). *Valtiomies ja Sotapäällikkö*. Otava: Helsinki.
Wavell, A. (1953). *Soldiers and Soldiering or Epithets of War*. London: Jonathan Cape.
Wei Liao-Tzu. (1993). *In The Seven Military Classics of Ancient China* (pp. 225–276). Ed. R. D. Sawyer. Boulder: Westview Press.
Yuen, D. M. C. (2014). *Deciphering Sun Tzu - How to Read the Art of War*. Oxford: Oxford University Press.
Zukov, G. Z. (1969). *Marsalkka Zukovin muistelmat*. Porvoo: WSOY.

CHAPTER 3

Time and Activity—Controlling Tempo and Seizing Moments

The use of the asset of time and temporality can be conceived of in fundamentally different ways depending upon operational needs and the commander's intent. Whether one chooses the offensive or defensive dictates how time is to be managed. For the attacker it generally is beneficial to win time and to spend as little time fighting as possible. He has the initiative and is thus able to dictate the time and the place of combat. For the defender it is just as beneficial to win time, but he wins it by spending it. That is, the aim of the defender is to consume as much time as possible from the enemy. But to remain on the defensive is a passive choice and cannot bring victory. As Vego (2009, p. V–49) wrote, "The offensive is a decisive form of war. Even when on defense, one's forces would be required to conduct offensive operations in order to obtain and maintain the initiative." Thus the defender must continuously be on the lookout for the right moment to act and then seize the initiative and go on the offensive. Activity and passivity, tactics of maneuver and attrition are just two sides of the same coin and it is dependent on the circumstances which one is more beneficial to be employed for certain duration of time.

Whether one chooses to be slow or rapid in his actions, speed must be managed and controlled properly to win time for one or to rob the enemy of the time at his disposal. Constant and unthinking acceleration of speed is not the answer to being faster since speed must remain controllable. Similarly mobility and movement for the sake of movement itself does not bring any gains. It must be purposeful. Time is relative and winning it is a

J. Hanska, *War of Time*,
https://doi.org/10.1007/978-3-030-45517-0_3

zero-sum game in which every activity must relate to what the enemy is doing at the same time. Instead of seeking to be faster and faster, doing things with variable pace works as a means to confuse the enemy and disrupt his timing. One must find the most suitable pace for his actions and vary this pace according to the rhythm of one's own preference and to the rhythm of the battle or operations itself. Instead of speeding up, one must be able to slow down or even stop according to the rhythm of operations.

3.1 Time in Offensive and Defensive Stances—Disturbing the Equilibrium

> As concentric operations are most advantageous in attacking, eccentric ones must, of course, be so in defence: every thing must be contrary in the two modes of war so opposite in their nature and their interests. (von Bülow 2013, p. 80)

As Bülow argued, defense and offense are juxtaposed ways of waging war, this must be reflected in the role time and timing play in them as well. The suitable approach to time management is relative and depends on the activity and behavior of the enemy. On occasion both armies attempt to reach the same objective irrespective of each other, such as occupying a certain area. If, however, their objectives are contradictory, likely the means of fulfilling their tasks also differ. As Leo VI (2010, p. 575) put it, "when the enemy acts boldly, entice him into premature, reckless action and useless maneuvers. If he is on the timid side, hit him hard with constant and rapid attacks. You must know the disposition of the enemy general and employ your own stratagems accordingly." Combat is a struggle of two opposing wills and if the enemy wants to win time, to slow him down is a priority.

To start with, one must first determine what kind of war will be fought and for what purpose. "War once decided upon, the first point to be decided is, whether it shall be offensive or defensive" (de Jomini 1992, p. 72; Handel 2001, p. 14). This seems self-evident, but the decision is crucial when it comes to the meaning of time in operational art. As Tukhachevsky framed the question, "Do we nip at the enemy because we are afraid of spilling too much blood, and thus seek to demoralize him? Or do we take him on, with the ultimate aim of destroying him?" (Cited in Simpkin 1987, p. 85). Generally defensive warfare requires one to waste enemy time by prolonging the war and wearing him out. In an offensive war one wants to win time from the enemy to bring the war to its conclusion as soon as possible. By wasting enemy time one can strengthen his defense or receive outside help. In winning time by acting rapidly one takes advantage of the enemy having less time for his actions. Nevertheless, even if one has chosen to partake in a defensive war, "the best thing for an army standing on the defensive is to know how to take the offensive at a proper time, and to take it" (de Jomini 1992, p. 183). It is a question of when to seize the initiative and switch from passive defense into the offensive. The timing is as important as the action itself.

The choice between offensive or defensive is related to time because in different phases of war the capabilities of the armies vary. As Triandafillov (1994, p. 51) wrote, "during the first months of a war, forces will be better prepared for those forms of combat not requiring complex shifts or maneuvers. These forces will be stronger in the defense than in an offensive." When forces consist of non-professional soldiers they are likely to have insufficient experience at the beginning of the war and receive their military training on the battlefields. Defense offers security and is likely to cause less casualties than offensive. Since it is also the stronger form of battle, it is a fitting choice for inexperienced troops. But prowess develops in the course of war and more complex operations become viable and once the commander is assured of adequate skills of his troops, the time to seize initiative and switch to attack must be exploited.

Military theorists often discuss attack and defense as mutually exclusive positions. It is easy to view the defender as the passive combatant and the assailant as the more energetic and active combatant (von Bernhardi 1914, p. 140). Seizing initiative is a highly praised decision in operational art and the initiative is always on the side of the attacker. This leads to such enthusiastic claims as "basic principles of our military preparation, of our operational art, are the principles of the offensive" (Isserson 2013, p. 42). Nevertheless, often even the commanders-in-chief are slaves to circumstances and the final choice is not theirs. "Treatises upon this subject sound for the most part as though attack and defence were exclusively a matter of free choice on the part of the combatants, whereas in reality this is hardly ever the case" (von der Goltz 1906, p. 151). Either their political masters or the situation itself dictates whether operational artists are allowed to attack or forced to remain on the defensive. Commanders attempt to make the best of the situation and utilize all their knowledge of operational art and exert their energy for the best possible outcome.

Offensive and defensive operations treat time in juxtaposed ways and this has not changed fundamentally even if technologies and doctrines have evolved. The proponents of network-centric warfare argued that the principle of offensive is "to act rather than react and to dictate the time, place, purpose, scope, intensity, and pace of operations. This is all about battlespace awareness, speed of command, and responsiveness" (Alberts et al. 2000, p.7–8). All the factors above need to be considered when one chooses to undertake an offensive battle. Should one adopt the defensive stance, he leaves initiative and the ability to dictate every one of these factors to the attacker. To defend is to remain passive and the best way the passive one can utilize time is to prolong the battle, the campaign, and the entire war. According to Strachan's (2013, p. 55) interpretation of Clausewitz, time is a great asset enjoyed by the defender, who can trade space into time while the offensive needs to be speedy in its execution.

The offensive approach is valued because military culture endorses activity in all circumstances. Even in WW I there was no doubt about the defense being stronger than offense due to weapons and prevailing tactics. If pure defense sufficed in warfare, as Echevarria (2000, p. 6) put it, "no attack equals no war equals no problem." The biggest challenge for defense is that "the defender is only victorious when he wins at all points, while the attacker triumphs if he gains the upper hand in a single spot" (von der Goltz 1906, p. 156). The principle of concentration of force favors the attack, and to win the defender cannot allow himself to be beaten anywhere. Mobile reserves provide the defender the opportunity to alter the balance of forces in his favor, but the attacker has the easier task to concentrate force to win at the single point of his own choosing.

Mobility is often wrongly perceived to be a tool of only the attacker. The defender benefits from mobility when he brings up reserves, counterattacks, switches to offensive, or when even a momentary slackening of offensive maneuver occurs. "Each pause in the movement was a benefit to the defensive. Defense forces could find time to dig in and brace themselves against heavy attacks" (Von Leeb 1991, p. 83). It is to the attacker's advantage to uphold continuous movement operationally and tactically to rob time from defensive preparations. The defender uses mobility to his advantage on the operational level by shifting troop concentrations in his defensive formation.

Traditionally defense as war of position is the strongest tactical option but with mechanized mobility, attack and defense go hand in hand on operational level. The same troops should not simultaneously employ both methods, but units should be used in turn and successively for both purposes. The task of the reserve units in defensive battle is not to defend, but to attack. The more mobile they are, the better their chances to accomplish their task in time. Furthermore, mobility of the defender must be succinct to counter that of the offense. "Rapidity and maneuverability in defense must correspond to rapidity and maneuverability in attack" (Von Leeb 1991, p. 121). The question is how should they correspond? For this discussion, WW I as the ultimate war of position is a suitable starting point. As De Gaulle depicted the essence of the war,

> The rigidity of the whole, and, as a consequence, the maintenance of the line, became indispensable. To wage battle 'in combined strength' was the strict axiom. There would have been no rest for the opposing forces had not their wings rested on impassable obstacles—Switzerland and the sea. Consolidation of the combatants in fortified positions did not change, but rather reinforced, the principle of the continuous front. If a local offensive succeeded in breaking through all the efforts of the defenders were devoted to re-establishing themselves shoulder to shoulder. Hinging, bracing, welding, warping, strategic withdrawal, these were the master words of military art. Meanwhile the attacker, who by his very success had uncovered his defences, slowed down his speed the further he advanced, and spoke only of pivoting, shortening the line, widening breaches, encirclement, reciprocal support. Up to the last shot fired, the opponents formed two flexible, but never broken, lines. (De Gaulle 1976, p. 127)

Perhaps it was natural development for a war that started with an attempt to execute von Schlieffen's version of modern-day Cannae, by outflanking the enemy, that flanking movements were to be averted by the defender at all costs (Ritter 1958, p. 48). And the best way to protect one's flanks is not to leave any. This irrational rationale led to fronts stretching across the entire continent. As Fuller (1920, p. 14–15) described it, "before the end of 1914, except for a few small breaks, a man could have walked by trench, had he wished to, from Nieuport almost into Switzerland." Extension of the fronts ensured that there was no way to outmaneuver the enemy by advancing to his flanks or his rear (Vego 2009, p. III–7). Thus the principal means available for the commanders were to perform local offensives, hammering into the enemy (Beaufre 1965, p. 64). The defender in turn did everything to keep the line intact. Even if the attack progressed, its momentum dwindled the deeper the penetration burrowed. Due to 'strategic withdrawals' the frontlines fluctuated temporarily, but in the big picture they remained almost immobile.

There is a difference between commanding offensive and defensive operations. In positional warfare everything is planned long in advance but in offensive warfare the commander is constantly making decisions (Ludendorff 1919, p. 37). Looking at military history, it is the great conquerors we remember and not the sturdy defenders. Offense is the trial by fire of generalship and when defense dominates the battlefield there always occurs a deterioration of operational art as well. This happened in WW I when "trench warfare was inevitably the negation of generalship, the triumph of mud over mind" (Liddell Hart 1927, p. 6). In military theory the preponderance of offense dominated while technology made defense immensely stronger. Everyone "suffered from this abuse of a correct idea, that of the offensive, because it was applied without discernment" (Foch 1931, p. lix). All wanted to attack, but defending was easier. This had its effects on mobility. As Liddell Hart (1927, p. 3) argued, "military history reveals that mobility has yielded to stagnation whenever the means of defence have acquired a material preponderance over the means of offense." Prior to WW I all armies were imbued with the spirit of the offensive. Attacking was the heart and soul of the military art. Fuller (1926, p. 258) wrote that there was no understanding of

> the protective power of weapons, and the result was static warfare. In the next war, if we do not realize the influence of new forms of movement on weapons and protection, the war, in place of being in nature static, will be dynamic in the extreme; we shall be swept into the sea or into some neutral country.

Fuller was right in his prediction. The British troops of WW II were literally pushed into the sea from continental Europe. The beginning of the war was devastating for the same reasons: not understanding how the means and ways of fighting had been altered. However, just as offensive means were re-invented during the course of WW I, new and inventive means were discovered in WW II to curb the mobility and the power of the offensive.

The relationship of defensive and offensive measures in operational art should resemble balanced scales, but often a seesaw is more descriptive. There is a tendency of warfare to search for balance. Measures to attain it are developed astonishingly rapidly, if the disequilibrium is profound enough and the need for restoring balance dire. Thus, in the course of a few years, from the dominance of the machine gun, the tank was invented to counter its defensive potential. Then, again, anti-tank measures and minefields countered the offensive, and equilibrium was restored. Operational artists need to invent measures to disrupt the status quo between defensive and offensive means in their favor and inflict operational surprises on the enemy. Again,

> a novel weapon or means of warfare, like an unknown plague, fills the imagination of man with horror and intangible fear. Yet, no remedy to this is to be obtained by locking up terror in a mental dungeon; in place, the unknown must be examined in broad daylight, its nature diagnosed and its antidote discovered. (Fuller 1923, p. 100)

The novelty of the new weapons and tactics has the potential to cause a temporary operational paralysis. As von Seeckt (1930, p. 54) argued, any technological offensive means have always been countered by measures employing the same technology. The ability to attack the imagination and create irrational fear is temporary since experience breeds familiarization. While the defender must employ all of his mental capabilities to develop an "antidote" the attacker has a limited period of time to make the most of his temporary advantage. Time is quickly running out and this requires a surprising and massive use of new weapons. Swinton wrote that because

> the chances of success of an attack by tanks lies almost entirely in its novelty, and in the element of surprise, it is obvious that no repetition of it will have same opportunity of succeeding as the first unexpected effort. It follows therefore, that these machines should not be used in driblets (for instance, as they may be produced), but that the fact of their existence should be kept as secret as possible until the whole are ready to be launched, together with the infantry assault, in one great combined operation. (Cited in Fuller 1920, p. 51)

Warfare is a competition between the man and the material. A sword is countered with a shield, a tank with a bazooka, and this cycle will keep turning as long as wars exist. Technology works for both the attacker and the defender and any offensive weapon has a dominance only until the defense gets accustomed to it (von Seeckt 1930, p. 65). Material and machines have not, according to Seeckt, gotten the best of man. They have developed to dominate the mass of men but not the man, since only he could operate the machine. In WW I the almost immobile mass of infantrymen was set against a mass of materiel and the bigger the mass of men becomes, the more certain is the victory of materiel mass over it (von Seeckt 1930, p. 65). The difference between the two World Wars is one of

> space and speed rather than of weapons. The weapon development favourable to the offensive has been counter-balanced by the weapon development favourable to the defensive. But other conditions have made the present war less static. While defence is stronger, space has been wider and forces faster. These offsets have given the offensive a better chance strategically than in the last war—by providing the attacker with no more room for manoeuvre, and more speed of manoeuvre, thus making it easier for him to achieve penetration. (Liddell Hart 1946, p. 22–23)

WW II lasted only a few years, but during that time the character of conflict changed profoundly. *Blitzkrieg* of the beginning turned out to be a quick thunderstorm, followed by heavy rain that bogged the mobility of the troops. Winning time by being fast metamorphosed into winning time by slowing down the enemy. Thus, the pendulum of war swings back, and we return to the foot soldier, not infantry trained to attack, but trained to defend. Not infantry armed with rifles and bayonets, but engineers equipped with anti-tank weapons. Finally we arrive at this somewhat perplexing conclusion: as the defensive gains on the offensive, as it always has after some new offensive weapon has been invented, and eventually 'bunker' it, military operations will become slower and slower, until battles between mechanized armies are likely to grow as static as they were between the enormous muscular armies of the World War (Fuller 1932, p. 290).

The defensive once again proved its strength over the offensive and the fast movement that robbed the defender of time was effectively slowed down until the defender was able to match it. In the course of time, offensive inventions hasten the pace of warfare and once the defender finds a counter-measure, the battles again become positional. Liddell Hart prophesized between the World Wars what the next war would be like. He actually gave two options for further development,

> the next war must either be more mobile than the last, which attained in its middle period the zenith of immobility, or it will cease to be war and become a mere state of impotent exasperation. The motive underlying all subsequent invention, or rather their military development, has been to dissipate

> this condition of stagnation, and every recent advance in military materiel has been in the direction of greater power and speed of movement. (Liddell Hart 1927, p. 128)

While all development prior to WW II aimed at increasing maneuverability and *Blitzkrieg* was an ode to mobility, in the course of the war this ode turned into a requiem. Stalingrad was a monument for the "impotent exasperation" Liddell Hart wrote about and generally bold maneuvers became exceptions to the rule the longer the war continued. As Svechin (1992, p. 255) stated, "It is easy to get involved in positional warfare, even against one's will, but it is not so easy to get out of it; no one managed to do it the World War."

3.2 Cycle from Passive Attrition to Active Maneuver

> In general, static elements of defense can be used to pin down, turn, or block the enemy's attacking force and thereby gain time for other friendly forces. Mobile elements can be constantly on the move and thereby prevent the enemy from exploiting his combat success or confuse him. (Vego 2009, p. V–55)

In other words, the mobile and static troops play different roles in operations and provide commanders with different capabilities to choose from depending on the situation. There are different ways of implementing operational art such as maneuver warfare, annihilation warfare, and attrition warfare. All of these can be used to fight both offensive and defensive battles. The last one is generally fought out to wear down and exhaust the enemy. Besieging and starving cities and fortifications was its classical form. Starvation and deprivation of necessities brought victory to the one with more endurance. Sieges, however, became obsolete during Napoleonic Wars. As mobility allowed operational art to re-emerge in the guise on maneuver warfare, siege warfare faded into obscurity (van Creveld 2011, p. 23). Positional battles of attrition didn't disappear but got new forms. They became 'material battles' and Svechin sarcastically depicted them as battles "designed to last entire months and involve trampling over the same piece of ground in an organized way and in which therefore gaining ground is less important than inflicting greater losses on the enemy than the losses we bear" (Svechin 1992, p. 273). This harsh judgment is not depictive of all battles of attrition. The discussion on the pros and cons of attrition and maneuver is included here because

> Time is perhaps the key discriminator between maneuver and attrition theory. Maneuver warfare is an intense contest for time. This has always been the case for those armies in history that have been disposed to fight according to maneuver theory (…) But the urgency of the contest for time is even more evident in modern warfare. The reason is that time is of greater value today than it was in the past. (Leonhard 1991, p. 82)

The theorists of attrition and maneuver seem to be portrayed in literature as crude stereotypes and opposites of each other. One wants to defend, the other to attack. The main difference is that the attritionist wishes to consume as much time as possible and the maneuverist attempts not to waste a moment. In operational art maneuver and attrition warfare blend seamlessly into different phases of an operation. The attacker may be forced into a battle of attrition. The defender wishes to diminish the force of the attack before switching into offensive and maneuver. This is because

> the object of all war, the crushing of the enemy's forces, can, after all, only be achieved by attack. The partisans of defensive also always maintain that it can only be assumed for a time, that in the end the defender must also begin to attack, and answer the thrust that he has parried by a stroke in return, and that he must ever keep this in view. That is to say, in other words, that the defender would also be the attacker, and only awaits the moment when he will be able to take the offensive with prospect of success. To make war means to attack. (von der Goltz 1906, p. 156)

As in all aspects of human experience there is no black or white but different shades of gray. Attrition and maneuver warfare are interrelated. As defensive attrition may turn into offensive with maneuver. Irregular warfare is often focused on attrition in the early phases while attempting to create a strong enough force to maneuver with in concentrated offensives. Warfare represents a continuous and repetitive cycle in which different methods characteristic to different times blend and follow each other as time and circumstances dictate. Different methods of operational art are never independent and self-sufficient. Simpkin (1985, p. 319) argued that there are

> three theories of war—attrition theory, manoeuvre theory and the doctrine of revolutionary war. All these lie on a continuum; and attrition theory becomes complementary to the others once fighting between organized and/or irregular forces breaks out. But just as, within manoeuvre theory, offence and defence are opposite aspects of the same continuum, so, within the continuum of the threat and use of armed force, attrition warfare lies at one pole, while manoeuvre warfare and revolutionary warfare adjoin each other astride the opposite pole.

A common trait among the great captains of history is their ability to use variability in operational art. The tendency to prefer maneuver over attrition comes and goes in cycles and often novel methods or new technological innovations rejuvenate maneuver. Marlborough became one of the early modern masters of mobile warfare not only because of his personal energy but also because of the growth of firepower, and flexibility of maneuver that had been lost for centuries (Colby 1943, p. 17). Frederick the Great was both a shining and a warning example of maneuver warfare. His art of war vacillated between bold movements designed to destroy the enemy in decisive battles and avoidance of battle through sieges and marches planned to exhaust the enemy. In a sense they were attrition warfare as well. They didn't drain the enemy of the lives of his soldiers but of his energies and supplies. Frederick and Gustavus Adolphus were skilled in choosing the right approach depending on the circumstances. There was "a wavering between the old and new systems, not did the genius of Gustavus Adolphus decide for either. We find no trace of a regular system in that memorable war" (von Bülow 2013, p. 235). Frederick was more systematic (Bond 1998, p. 17–18). The longer the Seven Years' War lasted, "the further the king moved away from the pole of battle and closer to the pole of attrition" (Delbrück 1997, p. 127). Frederick attempted to make the best of the situation and the assets at his disposal at any given time (Bond 1998, p. 24). If the army is not strong enough, a decisive battle doesn't bring victory and the genius of Frederick was in selecting varied means and methods for each occasion (Colby 1943, p. 103).

In most cases "of two opponents equal in other respects, the more active will be the victor" (von der Goltz 1906, p. 154). Activity in operational art is represented by the ability and willingness to maneuver. Theoretically maneuver is the optimal way of attacking, because it can fully utilize the mobility potential of the forces. At the same time defense is a static way of war in which attrition exhausts both forces in immobile frontal contact. Nothing could be less illustrative of the nature of warfare as a dynamic clash of forces and wills of their commanders. There can be no maneuver without attrition and vice versa. Without restrictions on mobility, movement becomes meaningless and it is excessively difficult to actually engage the enemy. For a mobile attack to succeed, a part of the forces need to engage the enemy, and temporarily tie him down and restrict his mobility. When the enemy is stationary, one can attack. A static element is a necessity in mobile warfare and static defense cannot bring a result without a mobile element that attacks once attrition changes power-ratios. Mobile battles and those of attrition are two sides of a coin in operational art.

Kaldor (2007, p. 24) argued that the theories of attrition as wearing down the enemy by imposing on him a higher casualty rate and maneuver theory as using surprise, pre-emption, mobility, and dispersion to create uncertainty in the enemy were concepts initially developed by Clausewitz. Clausewitz indeed wrote down the idea in a concise manner, but these methods have been employed as long as wars have been waged. Throughout history when the art of war is of exceptionally high quality, operational art favors maneuver. When troops get bogged down, the ugly head of attrition rears as happened in the aftermath of the industrial revolution. Mass-produced bloodbaths drained armies of soldiers and states of resources. For Ludendorff (1919, p. 176) the German offensive in Verdun failed to achieve a decision and turned into the first battle of attrition of WW I in which massive amounts of men and materiel were spent in unsuccessful attempts in the same locale while time dragged on. The breaking point of the wave of attrition warfare was reached in British offensive lasting throughout the summer and fall of 1917 in Flanders and summarized the principle of attrition warfare, "the concept of the battle of materiel is to defeat the opponent by crushing material superiority without relying on generalship" (Ehrfurth 1991, p. 523). If the state had enough manpower to train soldiers, agricultural capacity to feed them, and factories to supply them, an indust-real war proceeded like a clockwork toy. Operational art atrophied because it became secondary in importance to materiel.

Using Verdun as an example, Ludendorff (1919, p. 200) noted how to conduct a war of position and attrition. If one has chosen the offensive, it should be called off immediately when combat starts to turn into a battle of attrition, progress stops, and gaining ground by re-seizing initiative seems unlikely. When gains no longer balance the losses suffered, the attack should pause. The attacker can call off the offensive at any time. On the other hand, for the defender the battle of attrition continues as long as the offensive persists. Nevertheless, it isn't inherent that one would benefit from stubbornly remaining on the defensive. It is a question of rates of attrition and whom they favor over time. According to Guderian (1992, p. 37), a tremendous failure occurred in not realizing this and "time showed that the Germans suffered more than the enemy from the way they held on to positions that had been dictated by the needs of the moment, regardless of whether they were suited for defence over a period of time." The defender has to timely let go of the positions that have become too dear to hold. Even though

> an attacker may bleed to death before an adequately defended front, any attempt to hold one which can at best be manned on the scale of a safety screen will merely cause the meager defending forces to be expended at an excessive rate. Assuming, that is, that the enemy does not simply over-run them. (von Manstein 1982, p. 495)

Attrition warfare should not even be attempted without a certainty that the defending forces are strong enough not to be overrun before the goals of attrition are reached. As long as the attrition rate of the attacker is higher compared to the defender, passing of time favors the latter. Attrition should be chosen as only a short-term method.

> Joseph de Maistre wrote: 'A battle lost is a battle one thinks one has lost; for,' he added, 'a battle cannot be lost physically.' Therefore, it can only be lost morally. But then, it is also morally that a battle is won, and we may extend the aphorism by saying: A battle won, is a battle in which one will not confess oneself beaten. (Foch 1920, p. 286)

Foch's fantasies were part of an inspiring lecture to young officers, and for example Fuller shared this idea in early phases of his career (Trythall 1977, p. 27). Foch followed a long tradition in art of war to emphasize the moral over the physical as an element of heroic warfare. von Bülow (2013, p. 186) had earlier written that "not to be really beaten, you have only to believe that you are not so." Moral victories are inconsequential if physical conditions dictate otherwise. In terms of fighting spirit one may indeed not be beaten until one admits it, but an army can sustain only so many losses before yielding. During indust-reality the lethal potential of warfare ensured that one indeed could be destroyed even if one didn't confess defeat. Foch's thinking was characteristic to the First Wave but it has made a comeback in Third Wave warfare. This led Smith (2008, p. 19) to argue that "our conflicts tend to be timeless, since we are seeking a condition, which then must be maintained until an agreement on a definitive outcome, which may take years or decades." No doubt the coalition forces won a victory over the Taliban and Iraqi forces, but the wars dragged on regardless since the enemy did not admit losing the war.

The Peloponnesian War, of which the first part lasted ten years, (Thucydides 1971, p. 321–323) or the Thirty Years' War fit the pattern of fighting as little as possible. There were only a few land battles during the twenty-seven years of the Peloponnesian War which led Delbrück (1990a, p. 135) to call it "war without decision, through simple attrition." This attrition was more a mental erosion of the will to fight than expenditure of physical forces and resources. In the case of the Thirty Years' War there were thirty-three battles, none of which leading to military resolution. Belligerents attempted to exhaust the enemy economically while somehow supplying one's own forces (Münkler 2005, p. 42–47). In a war of attrition casualties pile up over time and resources poured into the war effort erode the economy of the state slowly but certainly, but the moral breaking point is far removed in time. In such a war there could be "neither victory nor defeat, only a common loss. To-day we are suffering not only from exhaustion of the body, political and economic, but from exhaustion of the spirit" (Liddell Hart 1932, p. 18). In contrast to attrition strategies there have been times in military history when the art of maneuver has dominated operational and strategic thinking to such a degree that battles were shunned. When wars were waged with mercenary troops the entire military profession would suffer in battles of annihilation. In warfare between men who have made war their livelihood all belligerents prolong war and simultaneously mitigate its destructivity. The incentive on both sides was to save lives and this led to avoiding battles but conducting skilled maneuvers. (Fuller 1961, p. 16, 22)

Joly de Maizeroy summarized the art of maneuver warfare as "not only in knowing how to fight but even more in avoiding the fight, in selecting posts, in directing the marches so as to reach the goal without committing oneself, … so again as to decide to fight a battle only when it is deemed indispensable." (Cited in Foch 1920, p. 27). The logic is perverted. Massenbach regarded shunning battles in favor of maneuvers as the supreme form of military art (Foch 1920, p. 27), but wars without battles represents decadence in the art of war (Foertsch 1939, p. 29). In rare cases, argued Clausewitz, a bloodless victory could be achieved by maneuver, but only if one was willing and prepared to shed blood and to fight (Howard 1983, p. 43). In strategy, fighting is not required for victory, but operational art is different. If warfare deteriorates to maneuver for maneuver's sake, there is no operational art, no victory. Combat decides wars and as Procopius advised

> in war one should not rely on luck and chance, even if one is very much stronger than the enemy, but should preferably seek to lie in wait for the enemy by means of ruses and stratagems. He who moves directly toward danger is not at all sure of victory. (Cited in Delbrück 1990b, p. 381)

The idea of avoiding open battles at all costs contradicts the principles of war. While one shouldn't rush headlong at the enemy at first chance, operational art should not develop into avoidance of all combat. Successful operations have phases of attrition and maneuver and the switch from one to another occurs fluidly. Classical examples of maneuver warfare—like some campaigns of Scipio, Frederick, Justinian, or Caesar—were not freely chosen, but brought about by a lack of men, material, or overall strength of the armies. In his own words,

> Caesar had conceived hopes of ending an affair without an engagement, or without striking a blow, because he had cut off the enemy's supplies. Why should he hazard the loss of any of his men, even in a successful battle? Why should he expose soldiers to be wounded; who had deserved so well of him? Why, in short, should he tempt fortune? (Caesar 2010, p. 178–179)

When commanders doubted their ability to create decisive battles, they resorted to other, more discreet and limited means of engaging the enemy. If both sides were too weak, warfare became a parody of itself and started to resemble complicated drills where sweeping maneuvers were conducted across empty terrain with armies circling each other in constant movement. This type of elaborate military posturing turned wars into armed show-offs. The state of war could last for a long time and actual warfare was at best spasmodic. Such noted *condottieri* as Paolo Vitelli and Prospero Colonna stated that wars are won by industry and cunning instead of actual clash of arms (Fuller 1961, p. 16), but mere maneuver is insufficient for winning wars since "one cannot succeed in breaking down the will of the enemy by mere exhaustion, but only through a victorious battle" (Von

Leeb 1991, p. 67). According to Delbrück (1990a, p. 522–523) the only one who ever managed to gain victory and complete surrender of the enemy without more than a few moderate-sized skirmishes and no major battle was Caesar in the civil war. This is the one exception that proves the rule that maneuver alone cannot bring decision (Heuser 2010, p. 89–96).

Napoleon made a huge leap forward in the art of war. The eighteenth century operational art relied on maneuver deeply entwined with diplomacy to influence the will of the enemy. Napoleon chose to crush the enemy's main force and thus effect a collapse of the will of the enemy state (Smith 2008, p. 61). He changed the rules of the art of war profoundly.

> There was a change when Napoleon came on the scene. A Corsican, not a Frenchman; a supreme careerist, not a true patriot; he was unchecked in pursuing his ambitions by any sense of responsibility for the ultimate welfare of his country as apart from himself. If his dreams were boundless, he took short views, since his horizon was his own life-span. Time was always against him; and he needed quick results. Now, whatever be the difficulties of winning a war by a battle, it is the quickest means—if it can be achieved. Hence Napoleon's predisposition for this means. (Liddell Hart 1937, p. 292)

This is a fitting description of Napoleonic warfare since it was a war against time above all else. Napoleon perceived that he couldn't create an empire that would last for millennia. He had his own lifetime to work with and he fully understood the waning of his vigor and energy in every battle and campaign. Just because Napoleon felt the pressure of time, he created similar pressure in his own campaigns and attempted to compress time to make as much as possible out of it. Napoleon didn't allow time win his victories for him but pushed continuously for results and decisive battles.

The further back in history we look the more devastating and simultaneously decisive battles we find even if they are few and far between. Already by late nineteenth century it had become very rare to achieve decisive victories (Bond 1998, p. 4–5). The size and complexity of modern armies make them difficult to destroy utterly, but the destruction of the enemy was the aim for the great captains of the ancient and less 'civilized' times. Spartans, for all their stamina in battle, didn't pursue their enemies for long (Thucydides 1971, p. 353), but Alexander, Hannibal, or Caesar fought cruel battles in which the pursuit of the fleeing enemy was ruthlessly executed until the exhaustion of their own troops. As Delbrück (Delbrück 1990b, p. 381) phrased it, "the highest principle of these commanders was: defeat and destruction of the enemy" and they had no moral scruples with casualties on the way to decisive victories attained by pursuing and slaughtering the disorganized enemy. Battle of annihilation is the ultimate aim of maneuver warfare in its highest form.

Thus the battle of Cannae held a fascination over German commanders of the first half of the twentieth century. This was a result of Clausewitzian doctrine of destruction of the enemy honed to perfection by Moltke. There existed a tendency to attempt to recreate the conditions of Cannae to annihilate the enemy. Von Schlieffen (1936) considered a frontal attack to be avoided and he wanted to use the space in the depth of the enemy for free movement to destroy him through encirclement. Attacks against both flanks to envelop the enemy had to be conducted for the battles to be decisive (Vego 2009, p. V–16; Bond 1998, p. 91). This preference for deep, sweeping operations was resurrected in mechanized warfare and, as an example, Rommel (1953, p. 116) wrote to his wife describing a battle about to take place, "It's going to be a 'Cannae', modern style." In reproductions of this ancient battle Hannibal had won,

> a broad battle line advances against a narrower but generally deeper one. The overlapping flanks envelop the hostile flanks; the cavalry attacks the rear. Should the flanks be separated from the center, for some reason or other, it is not necessary to assemble them again in order to continue the enveloping maneuver, for they can be sent at once against the flank and rear. (von Schileffen 1936, p. 234)

Von Schlieffen argued that not only Hannibal but also Frederick, Moltke, and Napoleon had used this method of wide envelopment in their greatest victories. Ultimately he conceded that 'Cannaes' are rare, since for battle to result in annihilation, there needs to be a genius like Hannibal on one side and a Terentius Varro on the other. In his eagerness to push the attack, Terentius Varro allowed Hannibal's troops to engulf and surround his own (Fuller 1948, p. 44). While geniuses are rare creatures, 'Terentius Varros' have "existed during all periods of history. Thus it happened that, though no real Cannae, with the exception of Sedan, has been fought, there has been a whole series of nearly-annihilating battles, and these have always occurred at the turning points of history" (von Schileffen 1936, p. 238–239). Such a victory requires ignorance from the enemy general that allows catching him unawares. One must surprise and the other must be surprised to make a battle of annihilation possible (Ehrfurth 1991,

p. 387–388). What surprises a general depends upon his professional abilities. Jomini (2007, p. 31) quipped that "a general may attempt with a Mack as his antagonist what it would be madness to do with a Napoleon." Even Napoleon's nemesis Wellington wrote that there "never existed a man in that situation, in any time, in whose presence it was so little safe to make what is called a false movement" (Bassford et al. 2010, p. 224). A less gifted operational artist can be enticed into a potentially destructive battle, but even then annihilation is unlikely. Delbrück (1990a, p. 523) listed battles in which the enemy was completely destroyed in the end. These included along with Cannae the Roman army at Lake Trasimeno, the Prussian army in 1806, and three French armies in 1870–1871. Even if a battle of total annihilation is often sought, it is seldom realized.

When both sides, trusted in their potential, decide the war in an open battle with the bulk of the enemy forces, maneuver warfare peaked. Hannibal, Genghis Khan, Napoleon, Frederick (at times), Gustavus Adolphus, or Sherman used their mobility and skill of maneuver to force the enemy into a decisive battle. Genghis Khan stands above the rest of these great captains as the precursor of the principles of mechanized warfare. All of his troops were mounted and their

> mobility arose from perfect assimilation to their means of movement—they were bred in the saddle and hoped to die in it—and from ability to live sparsely in order to travel light. Each trooper had spare mounts and carried his own subsistence, so that the Mongol army was unencumbered by masses of transport. (Liddell Hart 1932, p. 65)

This is a form of mobile warfare modern armies cannot replicate. The agrarian character of the societies they fought helped along with Genghis Khan also the Goths, Huns, Vandals, Normans, and Tartars to overrun one enemy after other. Only the invention of gunpowder, standing armies, and cooperation between civilizations ultimately negated the nomadic art of war (de Jomini 2007, p. 123). Unlike mechanized armies, a horse army didn't require extensive lines of communication and supply because its ability to forage allowed it freedom of operations. Every man and horse in a Mongol army was both a fighter and a weapon. Similarly, Carolingian armies carried what they needed for sustenance and foraged the rest. This made the army mobile, but consequently small in numbers (Delbrück 1990c, p. 22–24). The mobile units of the past and present have, however, in common the ability to inflict serious damage to the fleeing enemy.

Mechanized mobile armies with their tactical air support brought back into the art of war the prospect of pursuit (Bond 1998, p. 9). Mobility has to be exploited since when a commander has scored a victory,

> it is generally wrong for him to be satisfied with too narrow a strategic aim. For that is the time to exploit success. It is during the pursuit, when the beaten enemy is still dispirited and disorganized, that most prisoners are made and most booty captured. Troops who on one day are flying in a wild panic to the rear, may, unless they are continually harried by the pursuer, very soon stand in battle again, freshly organized as fully effective fighting men. (Rommel 1953, p. 96)

Disorganization of the enemy must be exploited. When disunity reigns among the troops and command alike the enemy is not able to manage time. His freedom of action is limited and the only options are to attempt to flee rapidly or set up a ramshackle defense. If the pursuit falters, the enemy wins time to regroup. Thus, the pursuit should be an organic element or phase of operation. Liddell Hart emphasized pursuit as a means of operational victory in the mechanized era while he acknowledged that originally Napoleon systematized it as a tactical function (Danchev 1998, p. 100). When the Red Army pursued Wehrmacht on the Eastern Front in WW II,

> realising that the enemy would make use of the slightest delay in our advance to organize resistance, we did all we could to keep up the pace of the offensive. To this end we dispensed with all regrouping manoeuvres, which would have inevitably slowed down our advance, if only briefly. As our units pushed forward, the frontage shrank and we could have pulled out some forces to the second echelon. But it seemed a pity to waste time, so we merely contracted the battle formations of the forward echelon. (Rokossovsky 1970, p. 305)

The Soviets understood the importance of momentum in keeping the offensive in motion. There was no time waste and thus every secondary issue was given less attention and all energy was concentrated on upholding movement. Pausing once an objective has been reached is a grave operational mistake. Sometimes conditions for pursuit need to be specifically created. This may happen when the attacker has enveloped or encircled the defender. The enemy may create a local superiority of force and break out of encirclement. Failing this, he may either surrender or in some cases fight to the last man. The desire to win time led to leaving the enemy an escape route out of the encirclement. This can be a part of operational plan that aims at annihilation.

> The proverb: 'A bridge of gold should be made for the enemy,' in connection with his retreat, is followed religiously. This is false. On the contrary, the pursuit should be pushed to the limit. And the retreat which had appeared such a satisfactory solution will be turned into a rout. A detachment of ten thousand men can destroy an army of one hundred thousand in flight. Nothing inspires so much terror or occasions so much damage, for everything is lost. (De Saxe 1944, p. 121)

If the enemy is provided with an escape route, pursuing it in flight would lead to lesser casualties than trying to destroy it in encirclement. In other words, the "bridge of gold" was a great idea if it only seemed golden, but was used to lure the enemy into chaotic flight under planned pursuit. It was not intended to allow the enemy to evacuate his forces but to hasten their destruction. Once the enemy is on the run, "the tactical advantage must be seized. The fleeing enemy who cannot be destroyed today, may return to the battlefield with his power restored tomorrow" (Rommel 2003, p. 187). Infantrymen lack the stamina or speed to pursue the enemy beyond a certain limit. As Wellington described the case in Waterloo, "I continued the pursuit till long after dark, and then discontinued it only on account of the fatigue of the troops, who had been engaged during twelve hours, and because I found myself on the same road with Marshal Blücher, who assured me of his intention to follow the enemy throughout the night" (Cited in Bassford et al. 2010, p. 18–19). Sooner or later pursuit falters due to the physical strain of upholding the momentum. To prevent falling prey to a trap set by the fleeing enemy the pursuer has to maintain a higher rate of mobility than the pursued. Only then can the pursuit continue relentlessly, exert constant pressure on the enemy forces, and prevent them from reorganization. The enemy,

> should he be able to get some rest, he could regain his strength, draw in reinforcement, and regain his powers of resistance, even of attack. The longed-for rest should not have been granted. The Second Army should have pursued. During a long period of peace nothing much was heard of pursuits. It was known from many maneuvers that the beaten enemy was, after a lost battle, as fresh, enterprising, and dangerous as twenty-four hours earlier. (von Schileffen 1936, p. 142)

Pursuit, stripped to its core, is an attempt to annihilate the enemy. It is a distinctive phase of the operation. "The act of annihilation or pursuit is virtually a new attack requiring fresh troops and troops of a more mobile nature to those pursued" (Fuller 1923, p. 54). Foch (1931, p. xxxviii) agreed that pursuit is "a second operation, carefully organized beforehand [...] ready to be put to immediate effect." Mechanized forces can quickly be replaced with new troops to continue pursuit.

While it is impossible to predict the outlook of future warfare, the necessity to save lives creates a pressing need to rethink the tenets of war. One must still echo Corbett's (1999, p. 227) claim that "whatever the form of war, there is no likelihood of our ever going back to the old fallacy of attempting to decide wars by manoeuvres. All forms alike demand the use of battles." On the strategic level wars can be won without fighting. Once warfare commences, there can be no operational art without combat.

3.3 Moments of *Kairos* and Culmination—Switch from Defense to Offense

> Success in war depends on coup d'oeil, and on sensing the psychological moment in battle. At Austerlitz, had I attacked six hours earlier, I should have been lost. (Napoleon, cited in Fuller 1961, p. 46)

Napoleon was master of moments and by all accounts was skilled in initiating action at just the right time. Not every moment in warfare is endowed with equal importance. War consists of lengthy periods of waiting in inactivity for a short burst of action at the most promising "psychological moment." There are periods of uninterrupted *kronos*-time when time moves linearly but between these periods exist moments of *kairos*-time that must be seized. According to Sun-Tzu (1993, p. 185), "armies remain locked in a standoff for years to fight for victory on a single day." The tempos of attack and defense are fundamentally different since their temporal objectives are contradictory. One aims at a standoff, the other at rapid victory. In indust-real warfare, "the elements of the offensive are rapidity and vigour; those of the defensive, perseverance and tenacity" (von der Goltz 1906, p. 158). At the most auspicious moment the defender must change into the offensive, since passivity doesn't lead to victory. It is

> offensive form alone, be it resorted to at once or only after the defensive, can lead to results, and must therefore always be adopted—at least in the end. Any defensive battle must, then, end in an offensive action, in a thrust, in a successful counter-attack. (Foch 1920, p. 283)

If there is no switch to attack, the initial attacker just withdraws, recuperates, and renews his attack with a new ruse or stratagem calculated to enable a breakthrough. The battle is not victorious for the defender if the attacker withdraws. It is merely a pause in the attacker's operations and a time for him to reset the tempo. Since we have thus far let Liddell Hart and others criticize him freely, let us discuss how Clausewitz treated defense and offense and when the decision to discard one in favor of the other should occur. For Clausewitz a defensive strategy consisted of finding the right balance between waiting and parrying or the defense and offense (Howard 1983, p. 54). Initially defense is a good posture to take, since,

> it is easier to hold ground than take it. It follows that defense is easier than attack, assuming both sides have equal means. Just what is it that makes preservation and protection so much easier? It is the fact that time which is allowed to pass unused accumulates to the credit of the defender. He reaps where he did not sow. (von Clausewitz 1989, p. 357)

The accumulated time is an important factor in defense, since it is won from the attacker, but nevertheless entirely defensive battles should not be carried out because defense always

> has a negative object, it follows that it should be used only so long as weakness compels, and be abandoned as soon as we are strong enough to pursue a positive object. When one has used defensive measures successfully, a more favorable balance of strength is usually created; thus the natural course in war is to begin defensively and end by attacking. (von Clausewitz 1989, p. 358)

Defense can be used to gain time and to prepare to attack. "The essence of defense lies in parrying the attack. This in turn implies waiting, which for us is the main feature of defense and also its chief advantage" (von Clausewitz 1989, p. 379). Waiting wins time, but doesn't help gain anything else (von Bernhardi 1914, p. 140; Bassford 1994, p. 32). Time won must be used for a purpose, and to reap the benefits of waiting it has to be combined with activity. As von Clausewitz (1989, p. 380) argued, "waiting and acting […] are both essential parts of defense. Without the former, it would not be defense, without the latter, it would not be war." To remain constantly of the defensive posture implies submission to the energy and activity of the enemy (Sumida 2007, p. 170–171). To wage war, one must combine defense and offense, waiting and activity, and understand the "dynamic play of forces." (von Clausewitz 1989, p. 379)

> Defensive warfare is most often the choice of the weaker party and time won by temporizing brings about the possibility to gain new friends for him as well as weaken and divide his enemies. Time, then, is less likely to bring favour to the victor than to the vanquished. (von Clausewitz 1989, p. 597–598)

Passing time generally works for the defender, but "time can become a factor in the conqueror's strength as well; but only on condition that a counter-attack is longer possible, that no reversal in conceivable—when indeed the factor is no longer of value" (von Clausewitz 1989, p. 598). Fighting a purely defensive war can at best bring about a 'draw' as the result of the whole war (Handel 2001, p. 7–8). One needs to prepare for seizing the initiative.

> While we may have more time and can wait until the enemy is at his weakest, the assumption will remain that we shall have to take the initiative in the end […]. So long as the defender's strength increases every day while the attacker's diminishes, the absence of a decision is in the former's best interest; but if only because the effects of the general losses to which the defender has continually exposed himself are finally catching up with him, the point of culmination will necessarily be reached when the defender must make up his mind and act, when the advantages of waiting have been completely exhausted. (von Clausewitz 1989, p. 383)

Waiting has to either strengthen the defender or erode the attacker. If these requirements are no longer fulfilled, the defender must actively seek the auspicious moment to initiate his offensive. Clausewitz routinely talked in terms of polarities, setting two things in a dialectical relationship to one another, but his interest covered the whole spectrum between the poles. There were no fixed relationships among the theoretical elements in his operational art (Moran 2007, p. 95). Thus, after having followed the initial diminution and resulting defeat of Napoleon's army in Russia von

Clausewitz (1989, p. 220) argued that the defender often is most powerful "in the heart of one's own country, when the enemy's offensive power is exhausted, and the defensive can then switch with enormous energy to the offensive." Once the initial fury of the attack has withered, one can take the initiative and commence offensive.

For Napoleon (1987, p. 62) "the transition from the defensive to the offensive is one of the most delicate operations in war." It is one of the *kairos*-moments in the course of a battle. Regaining the initiative may change the direction of the operation and create an opportunity to gain victory. One must choose wisely the moment to act, since an unfavorable one may accelerate the upcoming defeat. "At the commencement of a campaign, to advance or not to advance is a matter of grave consideration, but then once the offensive has been assumed, it must be sustained to the last extremity" (Napoleon 1987, p. 57). Napoleon viewed war as a zero-sum game. The actions one undertakes restrict the options open for the enemy. Making the decision to act must be thoroughly pondered, but once the die is cast, it must be followed through. de Jomini (1992, p. 22–23) argued that Napoleon and his campaigns give material worth studying for any aspiring general in many senses, since "he was sent into this world to teach generals and statesmen what they should avoid. His victories teach what may be accomplished by activity, boldness, and skill; his disasters, what might have been avoided by prudence." Activity and rapidity of movement are means of squeezing every ounce of advantage out of every moment, but failing to be prudent when necessary means taking the chance of losing all those advantages by ill-timed action.

Alexander had something no strategist, operational artist, or tactician of today is able to emulate. Even among active commanders, wishing to lead from the front, no one today is able to join battles like Alexander did. Delbrück (1990a, p. 231) called him a great commander on the battlefield but also in the grand manner because "he combined in one person the world-conquering strategist and the unexcelled courageous knightly combatant." Even if he was the greatest conqueror the world has known, he was simultaneously a warrior, leading his troops on the ground and participating in melees with a sword and a spear. The fact that commanders don't repeat these feats any more is not a fault, since Alexander's was a different time. In the words of Delbrück (ibid.), it was "the only moment in the development of warfare in which the elements of the conduct of war were so close to each other that the commander, following his nature, was at the same time a combatant." With all the risks involved, it speaks volumes of Alexander's warrior skills that he managed to keep alive long enough to establish his empire and simultaneously of his capabilities as a statesman that the demise of the empire occurred immediately after he died at a young age.

Tactical reserve made combining a commander and a warrior in the same person impossible. For Alexander leadership of battles was still relatively simple after tactical formations and method for battle were decided. When the order to engage was given, the commander could join his men in fighting because he couldn't control them after they were committed to combat. Reserves altered the situation and the commander moved from the front to the rear (Foertsch 1939, p. 55). As Delbrück (1990a, p. 232) wrote, "only with the advent of the principle of withheld units whose intervention as to time and place is directed by the commander himself is the latter's regular involvement in the fighting eliminated." To time the battle, to synchronize the use of his forces, and to commit the reserve to battle at the right time and the right place the commander's had to be distanced from the melee.

But what is the right time to commit the reserve? For Warden, the first rule of using reserves was that they should not be used piecemeal and the next two were "don't commit too soon, and don't commit too late." Since the right moment is not self-evident, determining it "must be a highly subjective process. It may even be a work of sheer artistry or genius" (Warden 2000, p. 101). It is instinctual to want to commit the reserves too early and it may be demanding to withhold them until the moment really is ripe. Surprisingly often in warfare "the real or supposed need of the moment often led to the forces being thrown into action too soon; sheer impatience sometimes led to miscalculations of this kind" (Guderian 1992, p. 43). If reserves are sent in at the wrong time, they cannot exert additional pressure but remain vulnerable to concentrated fire at the decisive point or they arrive too late. To have the forces converge in space but to fail to synchronize their effect in time reduces their effect. Right place but too soon doesn't allow for the utilization of the new forces. Too late, and the original forces may already be depleted. The time has to be *just right.*

Napoleon argued that precisely these special moments of *kairos*-time decide the winners of battles, and even a small action initiated at just the right moment can have a huge impact. According to him, "the fate of a battle is a question of a single moment, a single thought—the decisive moment arrives, the moral spark is kindled, and the smallest reserve force settles the argument" (Cited in Fuller 1960, p. 298). It is not a question of applying massive additional force but of exerting necessary amount of force in a timely manner to alter the balance. "There is a moment in engagements when the least manoeuvre is decisive and gives the victory; it is the one drop of water which makes the vessel run over." (Ibid.)

Every commander needs a reserve and this principle encompasses all levels of warfare from the smallest unit to the state itself. A tactical reserve determines the outcome of a battle, an operational reserve ensures operational success or consolidates it, and a strategic reserve is used to determine victory in a campaign or major operation with added importance (Vego 2009, p. VII–121). The reserve is ultimately a safeguard to ensure that the task will be successful. Reserves are "the principal tools in the hands of operational commanders for exercising and enlarging their freedom to act" (Vego 2009, p. VII–121). Reserves are a necessity for both the defender and the assailant alike, but whichever side employs them, the logic is the same. Reserves need to be "stationed in such a manner as to be capable of being transferred to a decisive point or direction in a minimum of time" (Sikorski 1943, p. 166). Their use may not become necessary, but they are the final trump card of the commander.

> The reserve is a club, prepared, organised, reserved, carefully maintained in view of carrying out the one act of battle from which a result is expected—the decisive attack. The reserve is spared with the utmost parsimony, so that the instrument may be as strong, the blow as violent as possible. (Foch 1920, p. 300)

The assailant must send his reserves to action when applying additional force has the best chance to turn the tide in his favor. The defender has a slightly different situation. Either the assailant can create a breakthrough and he has to commit reserves reactively to block it or he can make an active choice during an auspicious moment and switch into the offensive. Reserve is used to secure a decision and thus the counter-attack with reserves is one of the *kairos*-moments for the operational artist or tactician. When the reserves are committed, it must be done boldly. If reserves are committed piecemeal their effect diminishes. When reserves are thrown in, the battle or operation usually gets decided either way. The logic dictating employment of reserves has remained almost unaltered over time.

> Success in employing one's operational reserves essentially depends on the timing of their commitment; otherwise, the operational commander can miss the opportunity to obtain the initiative and also enhance freedom of action for the enemy commander. Equally dangerous is to employ one's operational reserve prematurely. One's premature actions are usually the result of haste and lack of organization. They also might be caused by one's unrealistic assessment of the enemy. In contrast, one's preemptive actions are planned and organized with the aim to achieve surprise. In short, both premature actions and actions taken too late are a good indicator of a commander's failure to use the factor of time properly. (Vego 2009, p. III–28)

Once again, the demands of proper timing are excruciatingly harsh on the commander. Thorough planning on the use of the reserve has to be conducted based on a sound estimation of enemy's capabilities. That is, the commander and his staff must be able to pre-determine with considerable precision the tempo of enemy action. Small mistakes in estimations can be rectified by creating a reserve with high mobility and ensuring it will be able to move at a moment's notice, but the margins are tight. The faster the battle, the more difficult the *kairos*-moment becomes to identify.

Finding the correct movement to seize the initiative is perhaps the foremost *kairos*-moment of operational art and a chance for the genius to shine. The importance of this moment can be found in every theorist from Sun-Tzu to Napoleon and even the doctrines of today. von Bülow (2013, p. 84) optimistically argued that it could happen at any time "by the simple act of falling upon the flanks of the enemy, and attacking his rear." Even if the act itself sounds simple, it is not easy to execute. The difficulty of identifying *kairos*-moments has increased as warfare has evolved and battlespace has extended into new dimensions. "In these days of great distances it is most difficult to recognize the turning points in the progress of battles" (von der Goltz 1906, p. 162). This is only partially true because not only the distances but the duration have grown and decisive moments are further and further apart in the flow of time. However, the more complex operations have become, the more such potentially promising moments to act there have been, but the harder they have become to identify. The operational artist cannot endlessly await *kairos*-moments. Sooner or later, and preferably sooner, he has to seize the moment and turn it into a time of *kairos* by decisive action. Space and time conjoin since the right moment is always related to both time and place, and instead of stationary, its location is fluid. Perfectly timed action is useless if the locus of the action is incorrect and vice versa. For an operational level commander,

> of all faults, only one is ignominious, inaction. We must therefore constantly try to create events instead of submitting to them, and to organise attack from the first, the rest being subordinate and having to be considered only from the standpoint of the advantages which may result from the attack. (Foch 1920, p. 284)

Foch's claim about being active in attempting to create events and not merely succumbing to them could be included among the principles of war. There are times when remaining passive is beneficial, but even then within the passivity must grow seeds of activity, ready to sprout forth at the opportune moment (von Bernhardi 1914, p. 158). If during combat time benefits the offense and weakens the defense, it is a priority to find or actively create the *kairos*-moment that promises a chance to turn the tide, since each passing minute makes it more probable that a more opportune moment will not present itself. It is when the attacker has

> taken the initiative in movement, the defender is naturally exposed to the danger of arriving at the critical point later than the former. It will thus be necessary for him to impede the movements of the enemy. This may be done by means of counter-attacks, which come upon the opponent in the midst of the execution of his designs. (von der Goltz 1906, p. 159)

When execution of one phase in battle or an operation has ended and the next is about to commence may be the most auspicious *kairos*-moment to re-seize initiative. Between successive phases of even a successful offensive operation some confusion is bound to reign and a counter-attack may turn it into full paralysis. Another potential *kairos*-moment occurs when

> before the strategical defensive changes to the offensive, a pause takes place. The original attacker is crippled and cannot go on. The defender has got rid of the feeling of being mastered, is strengthened in moral and physical respect, and the change of role is thus gradually brought about, long before it is actually announced, to the surprise of the observer not on the spot. (von der Goltz 1906, p. 162–163)

If the attacker is unable to uphold momentum, when his initial effort falters, there is a pause until new units replace the spent ones. This offers a *kairos*-moment when the first and second wave of troops are intermingled. This type of active defensive timing is suitable to counter the Soviet doctrine of deep battle or operations, but it can be employed in every scenario when the attacker runs out of momentum and reinforces his attack with additional troops. When and wherever the enemy is disorganized, the counter-attack has higher probability of success. The same idea was expressed in Chinese texts.

> When the enemy's ranks—front and rear—are not yet settled, strike into them.
> When their flags and pennants are in chaos, their men and horses frequently shifting about, then strike into them.
> When some of their officers and troops advance while others retreat; when some move to the left, others to the right, then strike into them.
> When their battle array is not yet solid, while their officers and troops are looking around at each other, then strike into them. (T'ai Kung 1993, p. 102)

While T'ai Kung talks of the time when the enemy is just creating his battle-formation, the ideas apply as well to the moment when the second wave of attack is about to commence. The same confusion reigns and the ranks are as yet unformed. Even if the first wave is not crippled by casualties, materiel is somewhat depleted and the soldiers weary. The second wave of troops is entering hitherto unknown territory and tread carefully to avoid friendly fire while responsibility for the front is exchanged. No matter how disorganized the enemy is, the time to attack is only when one's own troops are ready for it. One should commence the counter-attack in accordance with the Vietnamese principle: "strike to win, strike only when success is certain; if it is not, then don't strike" (Giap 1962, p. 169–170).

The right moment to switch from the defensive to the offensive should thus in general happen when the power balance between the two forces is on the brink of changing. The attacker can always dictate the time, concentrate forces, and through seizing the initiative master the moment. He concentrates his power for a breakthrough, and the longer it takes for him to create it, the more his combat power wanes. The culmination point of the defender is reached if the attacker manages to create a breakthrough or the ability to initiate a counter-offensive is lost. The attacker's culmination point in turn is reached when his concentrated power no longer exceeds that of the defender (Vego 2009, p. VII-75–78); Echevarria 2000, p. 15).

The point of culmination of the enemy attack is a moment for the defender to attack. The elusive nature of the point of culmination makes it extremely difficult to identify in combat. If culmination point is indeed "*an area of uncertainty or nonrecognition in terms of space and time*" (Vego 2009, p. VII–74) the commander on the defensive must look for a moment when the fierceness of the attack seems to abate. However, the enemy may be just re-adjusting his tempo or gathering momentum for new offensive. The balance of the respective combat powers of both sides is hard to quantify and the commander must rely on his *coup d'oeil* to find the *kairos*-moment. This can be the culminating point of the attack when one reaches

> the equilibrium of the combat power of both sides in a conflict. It is a point at which the attacker can still revert to defense to protect his gains and regenerate his combat power. When equilibrium exists, theoretically a determined and agile defender can take the advantage and go on the counterattack or counteroffensive. A culminating point in combat is reached when the relative combat power shifts between two sides. It is the point beyond which the difference in relative combat power begins to decrease rapidly. Theory does not say that if a commander continues beyond the culminating point he will face a defeat. It only warns that the risks of setback and even defeat will be pretty high if he does so. (Vego 2009, p. VII–74)

Even if theory and practice differ, this can be considered to be one of the most promising *kairos*-moments. If the equilibrium of combat power is unstable, it is likely that sudden change in defender's rhythm, pace, and activity may create disequilibrium and defeat the attack rapidly. Any moment when the attacker is in a state of mental uncertainty is a favorable moment for bold and decisive action of the defender. The importance of culminating point cannot be discarded. One must do his utmost to avoid reaching one's own culminating point and already in planning phase identify possible actions he might initiate to speed the enemy toward his.

3.4 Slowing Down to Manage Time and Tempo

> While quick decision is the general principle, we must oppose undue impatience. (Mao 1963, p. 142)

Mao waged a very protracted war in China which had seemingly inexhaustible strategic patience, but even in operational art the balance between being quick and being patient plays a role especially in maneuver warfare. We often focus on time won from the enemy by engaging him sooner than he expects, but mobile warfare can win us time in another way. A small army can employ its superior mobility to gain protection. A rigid line of defense remains easy to penetrate with excessive force, but if lines are mobile and reform fluidly, one can win time for defensive concentration of force and a counter-attack as the enemy is slowed down. High-mobility forces theoretically could evade the massed attack of the enemy. Armies require ability to use the resources of space and time to gain an edge. This ability can be divided into reliance

> first on its speed (above else on the rapidity with which the situation is clarified and understood), second on rapidity of decision (*Entschluss*), and finally on speed of execution. Delay on the part of the command must be made good by swiftness of movement. The commander must never wait for orders where he can gain success by independent decision. (von Moltke 1993, p. 161)

The more time is consumed in comprehending the situation and reaching a decision, the less time is left for the execution of the orders. This is why Moltke emphasized movement as timesaver. But as armies grew more mobile, the amount of time potentially spared by speed of movement became less important than time saved in decision-making. Instantaneous reaction to any situation is crucial.

> Every soldier knows that it is of vital importance to quicken the tempo of operations and the rapidity of manoeuvre. It is the way to gain the advantage in modern battle, and to seize the chance of exploiting any advantage gained. The proverb that "opportunities are fleeting" has an intensified truth since armies have become mobile through motor power. (Liddell Hart 1950, p. 290)

In contrast to indust-reality the Third Wave armies are facing a somewhat different challenge. The need to accelerate the tempo and maneuver is still pressing, but the issue to tackle today concerns control of the elevated speed. Operational artists need to find the maximum operable speed for the troops. This can be defined as the pace of operations that is as fast as possible to exploit mobility and surprise but simultaneously keeps the maneuver of troops and the pace of decision-making controllable. The problem with elevated tempo and rapid maneuvers is that instead of surprising the enemy, one may end up surprising himself in turn. The tempo has to suit the mobility and operational agility of the troops.

> Speed is dependent on a force's mobility. High tempo and movement at great depth can enhance the effect of mass. But higher tempo entails higher risk, because the shorter time does not allow for comprehensive planning. Whether the tempo is high or low depends on the force's mobility; its tactical rate of advance; the time to complete movement; quality of intelligence; patterns of combat support; quality of logistic support and sustainment; and so on. A low tempo allows more time to recover from one's errors or mistakes and to prepare a more comprehensive estimate of the operational situation. (Vego 2009, p. IX–150)

Low tempo is occasionally good, because it allows for preparations and may reduce the amount of mistakes. The early proponents of mechanized warfare had an urge to only increase the velocity of operational art. There was no mention of deceleration and adjusting speed to prevailing conditions, the most suitable rhythm for one's own forces, or the actions of the enemy.

> Every gain in speed increases not only the attacker's security but the defender's insecurity. For the higher the speed the greater the chance of, and scope for, surprise. Speed and surprise are not merely related; they are twins. (Liddell Hart 1932, p. 60)

During the Third Wave, speed favors the assailant only when he is able to control it and either accelerate or decelerate it at will. If speed is allowed to accumulate on its own, there comes a point when security evaporates. Velocity increases the chances of surprising the enemy, but with high speed, timely reactions to changes in the situation become impossible and the frequency of mistakes grows. There are risks involved with constant acceleration and the fetish for movement. "We live in an age of rapid and unthinking movement. The railway, the steamship, the motor-car, and the aeroplane whirl us from place to place" (Fuller 1932, p. 37). This characterization is depictive of the mechanized period of indust-reality but remains valid today during the Third Wave. Movement must be justified, planned, understood, and controlled. It should be "rapid" but not "unthinking."

Whether we are talking about moving troops or information around the battlespace, all movement must be imbued with a sense of purpose and its speed must be restricted to abide within the limits of control and comprehension. It is a tendency of military development to increase speed. Tommy Franks, for example, saw the second Gulf War to revolve around speed and argued that the way to win the war was "by getting inside the enemy's decision cycle. Remember: Speed kills … the enemy" (Franks 2004, p. 466). But speed can be lethal for one's own operations as well. It is generally a good thing to keep the decision-making cycle revolving rapidly. The purpose is to manage the cycle to compress the time at one's disposal and simultaneously disorient the enemy forcing him off-rhythm. Furthermore, within decision-making process multiple cycles like collection and analysis of information spin simultaneously. In complex organizations there are several loops on different levels and some are faster than others (Kagan 2006, p. 105). As Boyd described his OODA loop, "Note how orientation shapes observation, shapes decision, shapes action, and in turn is shaped by the feedback and other phenomena coming into our sensing or observing window" (Cited in Coram 2002, p. 344). Tempo and the ability to properly synchronize revolutions is important. While it is somewhat preposterous to compare operational art to playing chess because of the stakes involved, there is a similarity. Once the game starts, one defends, the other attacks; one retreats, the other advances (van Creveld 2008, p. 67–69).

> Every chess master knows and appreciates the value of tempo. Simply put, tempo consists of the player's pace of moves, such that the opponent, who may have a good plan, has no time to execute it. The player with tempo constantly forces the opponent to react defensively to a series of attacks, threats and feints, all the while advancing his own plan. He need not concern himself overmuch about enemy intentions, because his tempo serves as a shield against enemy attack. (Leonhard 1991, p. 16)

Influencing the enemy's freedom of action by enforcing one's own preferred tempo on him is crucial. Depending on the speed preferable to the enemy one should even prepare to slacken one's own tempo. There is an urge to seize chances rapidly and exploit advantages, but one can sometimes wait for those chances to actualize. Tempo must be adjusted so that the ratio of right decisions is not diminished. Liddell Hart (1950, p. 290) lamented that, "if the operating speed is faster that formerly it has not quickened to anything like the extent represented by the difference between the old marching-pace and modern motor-pace." This indicated that human intellect still remained in charge. The elevated pace is harmful, if commanders cannot make their decisions timed to the rhythm of war. During the Third Wave we must avoid letting computers or other machines set the rhythm for warfare. De Gaulle (1976, p. 47) wrote that,

> at present the machine controls our destiny. Certainly, from the dawn of history it had relived the toil of our fellowmen. [...]Throughout the ages, neither the part it played nor its form were modified. But the last century radically changed the relations between human beings and their mechanized servants.

During the boom of mechanized warfare machines increased the toil of the soldiers and commanders since everything was done according to timetables of the machines. Because tanks could travel at a certain speed, this became the requirement of warfare. Because machine was able to do this or that, the human had to aspire to the same. It was a maxim of mechanization that "an army superior in activity can always anticipate the motions of a less rapid enemy, and bring more men into action than they can at any given point, though inferior in number. This must generally prove decisive and ensure success" (Fuller 1946, p. 23). Undoubtedly activity is the key to mastering situations, but activity is also about being mentally able to grasp the intricacies of each emerging situation instantly. One needs to be fast in both physical and especially mental maneuvers. Although movement occasionally slows, thinking cannot. Patton (1947, p. 349) put the essence of time management into words quite beautifully by contrasting

> Haste and Speed: There is a great difference between these two words. Haste exists when troops are committed without proper reconnaissance, without the arrangement for proper supporting fire, and before every available man has been brought up. The result of such an attack will be to get the troops into action early, but to complete the action very slowly. Speed is acquired by making the necessary reconnaissance, providing proper artillery and other tactical support, including air supports, bringing up every man, and then launching the attack with a predetermined plan so that the time under fire will be reduced to the minimum. At the battalion level four hours spent in preparation for an attack will probably insure the time under fire not exceeding thirty minutes. One hour spent in the preparation of an attack will almost certainly insure time under fire lasting many hours with bloody casualties.

Making a hasty attack prolongs the battle, since despite the early start the following stages are going to be slow and proceed with a lot of friction. To act with haste is often to adhere to an insufficient plan. The commander's calculations must weigh the advantages derived from spending more time in preparation and planning and having to use less time in the execution of the plan in actual battle. To extend the planning phase often means shortening implementation. If the enemy is active and attempts to seize or regain the initiative, time-consuming detailed planning is disadvantageous. If the enemy is in inactive defensive posture, more time can be allocated to preparation.

Patton's arguments resonate in operational art. There is a need to save and manage time on every level of warfare. Yet, the meaning of time is different when one ponders the benefits and losses one can experience. The higher up time is managed, the more drastic are the consequences of failures and the more promising the successes. Generally, "the more senior the officer, the more time he has. Therefore, the senior should go forward to visit the junior rather than call the junior back to see him" (Patton 1947, p. 350). This time-management rule of thumb fits tactics and operational art, but its implications are to be found in strategy too. There is more time higher up the hierarchy while the importance of managing it is elevated as well. Thucydides (1971, p. 59) wrote that "if you take something before you are ready for it, hurry at the beginning will mean delay at the end." This strategic viewpoint applies to operational artists alike, but in tactics, time is always short. Things need to be done rapidly, sometimes even in haste, but their consequences are not so far-reaching. Thus, the task of each commander is to delegate time to his subordinates.

If time cannot be managed on the tactical level, battles may be lost, operations falter, and the strategic goals will not be reached in time. If, however, on operational level time is lost and plans, operation orders, and warning orders are not drafted early enough to allocate subordinates enough time for preparation, the execution may fail completely. While individual battles still may be won, they don't lead to profound results since inner cohesion of an operational plan doesn't direct, guide, and interlink them for the fulfillment of a common goal. Rommel had a clear-cut view on the importance of time management. One must just always be quicker than his enemy in both decision-making and execution of those decisions. He wrote that in mechanized warfare

> speed of operations and reaction time of command were decisive factors. The troops must be able to carry out their tasks with the greatest speed and without delay. One cannot in this situation be satisfied with the norm, but must constantly reach for and demand the best, for whoever makes the most effort is quickest—and the battle always goes to the quickest. (Rommel 2003, p. 133)

We can read between the lines that while Rommel's ingenuity as a tactician or an operational artist is unquestionable, his talents didn't fully extend to strategy. His writings treat war as a string of battles and not as a total exertion of the entire armed forces and the state. Commanders have different objectives and responsibilities on different levels and in tactics the purpose is

> to win battles. At the operational planning level, however, the battle is a building block, not an end in itself. The operational commander must be skilled at using battles (whether won, lost or avoided altogether) along with other assets (maneuver, deception, interdiction, etc.) to structure a winning campaign. Hence, operational art provides the justification for a battle. (Leonhard 1991, p. 11)

Rommel's excellence in operational art consisted of this skill of structuring a campaign out of the isolated battles fought in the desert. In operational art the temporal seamlessness in the operational structure is evidence of success. The two-fold acceleration of operations and reaction time in command enhances efficiency and wins battles. Rommel demanded an ever-quickening pace of operations. According to him, the commander should never be satisfied with and accept the performance of his units. There needs to be a constant increased effort to be faster than the enemy since this leads to increased speed and the quickest is the winner of any battle. Rommel recognized that his skills were in battle-command, since "time and time again I have learnt that in close-quarter combat the winner is always he who manages first to envelop his enemy with fire in the shortest time while the one who waits must be the loser" (Rommel 2003, p. 38–39). This is simple like a showdown in the Wild West; the first to shoot wins. Time has a very straightforward meaning. Just be quicker than the enemy and victory is yours. "You win or lose, live or die—and the difference is just an eyelash" (MacArthur 1964, p. 145).

The requirements for winning time tactically come from the temporization and timetables of operational command. To fulfill time-related operational demands junior commanders should be given a rough outline of the timetable for the campaign. "The sole criterion for a commander in carrying out a given operation must be the time he is allowed for it, and he must use all his powers of execution to fulfill the task within that time" (Rommel 1953, p. 119). Operational commanders time the operation and give junior commanders demands when to reach their objectives. Then they know the time at their disposal and can plan their actions accordingly. But the timetable of operations has become tighter and tighter.

> One must count on more frequent combat and, besides, the intervals between successively developing operations will be shorter than during the first period of the World War and where the Civil War is concerned, these periods of time are not even comparable. (Triandafillov 1994, p. 131)

As mechanization increasingly pervaded the troops, the pace of activity in warfare increased and changed the structure of an operation. Battles were no longer isolated and a constancy of combat replaced their earlier intermittent nature. Operations began to follow each other in close succession. This was a logical continuum from the recognition of the importance of speed in warfare. The race to make consecutive battles immediate sequels to the first one led to a situation where along the course of combat, operational plans needed to be readjusted. Fighting cannot continue unceasingly indefinitely. Periodically momentum runs out and is replaced by inactivity until new momentum can be built. Thus, time-manipulation requires ability to restore the energy, equipment, manpower, and morale of the troops rapidly to achieve battle-readiness before the enemy.

Since war is a reciprocal activity, it must be recognized as a general rule, that, "any slowing down of one's own operations tends to increase the speed of the enemy's. Since speed is one of the most important factors in motorized warfare, it is easy to see what effect this would have" (Rommel 1953, p. 285). Operational tempo can be slowed down only when the resulting increase of the speed of the enemy is unlikely to inflict serious harm. Adhering to the same logic, if one wants to slow the enemy down, he must increase his activity and pick up his own speed. Pre-mechanized forces were unable to

> create a centre of gravity quickly enough, and they thus lack the quality which matters most in mechanized warfare. Because of their lack of speed the enemy can take them on one after the other, each time with locally superior forces, and destroy them piecemeal without suffering undue casualties himself. (Rommel 1953, p. 379)

Superior mobility makes it operationally possible to create favorable force concentrations where and when required. Mobility enables successive attacks and creation of temporary superiority at chosen points, but the duration of this superiority has to be fully utilized. As Tukhachevsky described it, "the transition from break-in battle to turning movement must be carefully thought out and adequately planned. These offensive phases must follow one another without any gap in time, let-up in intensity, or hiatus in communications and resupply." (Simpkin 1987, p. 92)

Whether to slow down or increase the operational tempo depends on the objective one wishes to accomplish. Naturally the decision-making cycle has to ensure that incoming information is always processed so that time is not wasted in producing decisions. This is a requirement of all military operations and the idea of adhering to different tempos doesn't apply in the realm of decision-making. During the execution phase the pace of operations needs to be adapted to suit the fulfillment of objectives. "He who would benefit by gaining time and saving his forces must not of his own motion increase the tempo of the war. The weaker one is in war, the more he must seek to profit from the mistakes of the enemy" (von Freytag-Loringhoven 1991, p. 200). It is advantageous for the weaker side to slow down the speed of the enemy and the rhythm of the entire operation. Combined with the concept of pace with variability this is an asset. If the defender can impose a slower pace of operations on the attacker while preparing his counter-attack, the sudden change in tempo has added shock value. The drastic change of the tempo sets the enemy off rhythm and forces him to readjust. To alter the tempo of operations is to seize initiative.

3.5 Pausing to Win Time

> In general, one of the operational commander's most difficult tasks is to predict and identify the culminating point and whether it has been exceeded or reached. Hence, it is critically important that the operational commander timely and accurately sense or anticipate the approach to, or arrival at, the point of culmination during a campaign/major operation, so that the

> ultimate objective can be accomplished. Such a time- or force-sensitive decision creates a danger for a commander who cannot adequately balance ends and means, because it will result in a mismatch between combat and sustaining resources that may force the campaign or major operation to culminate before the assigned objective is reached. (Vego 2009, p. X–80)

The attack must never pause because it has reached the culminating point Vego discussed. Either the objectives must be set so that they can be reached before momentum runs out, or a carefully timed pause has to be pre-planned. There will always be pauses in operations even if modern theories seek to get rid of them. Some pauses are beneficial while others caused by enemy action at unplanned times are inherently harmful. As an example of a properly planned one we can use Napoleonic era and the pause between the approach march to the battlefield and the battle itself. Limited range of weapons made it possible to rest the soldiers during this pause (Isserson 2013, p. 16). During combat the situation was different. The objective could never be set beyond winning an individual battle and the entire battle had to be fought before the energy of the troops was exhausted. Win or lose, there had to be a pause before the next clash of forces. Troops had to rest.

> Rapidity and continuity of action are the elements of the attack. No halt may be thought of before the object has been attained. Any suspension of operations is dangerous on account of the reaction succeeding a period of unwonted activity. It is difficult, during the operations, to resume the offensive once it has been suspended; and to renew an attack in the course of a battle is practically impossible, unless reinforcements arrive and give a greater access of force than is lost during the period of relaxation. Hence the deliberate suspension of an attack is never justified. (von der Goltz 1906, p. 157)

The objective should be set meticulously. It should not be too far either in terms of distance or time to be reachable before momentum is depleted. Mechanized mobility enabled the commander to pull in fresh troops to continue the battle or conjoin several battles into an operation without a discernible pause in action. Once the first wave of troops was broken and their energy spent, the next wave could be employed to continue the advance. The pauses still occur for individual units and are just as necessary, but from the operational perspective the campaign continues without standstills. The objectives have to be planned so that units reach them in desired temporal succession, and mastering the moment when a unit continues the attack of the previous one requires detailed preparation and immaculate timing.

Jomini (1992, p. 114) argued how "it is impossible to sketch in advance the whole campaign. The objective point will be determined upon in advance, the general plan to be followed to attain it, and the first enterprise to be undertaken for this end: what is to follow will depend upon the result of this first operation and the new phases it may develop." War gains momentum by itself—or falls into stagnation. The direction of war is altered after every major battle. They can be regarded as moments of closure when the situation is evaluated and new directions are chosen if the operation so dictates. Clausewitz phrased the same idea differently,

> Since war contains a host of interactions since the whole series of engagements is, strictly speaking, linked together, since in every victory there is a culminating point beyond which lies the realm of losses and defeats—in view of all these intrinsic characteristics of war, we say there is only one result that counts: final victory. Until then, nothing is decided, nothing won, and nothing lost. (von Clausewitz 1989, p. 582)

On operational level when the battle is brought to closure, there lies a point of culmination which opens up new vistas beyond victory or defeat. Only then the next move can be determined. School books represent wars as pre-determined processes running toward their 'inevitable conclusions.' In fact, the historian emplots the war by joining respective activities of units into a logical whole by creating a narrative that presents somewhat illusory coherence of action. After every battle the war could have taken a different turn. The operational planner is tasked to emplot a campaign before it commences despite the impossibility to see beyond the initial clash. He has to take into account that the thrust cannot continue infinitely and plan a suitable moment to halt the advance. This is a part of time management. A preplanned pause can occur at the most suitable moment. One can plan for replenishment of ammunition, petrol and other supplies, a change of units at the point of the spearhead, and other administrative issues which hamper the operation unless they are timely conducted. No matter how pressing the urge to push deep into enemy territory, one has to evaluate the gains of pushing onward.

Occasionally commanders have chosen more straightforward approaches and Patton's (Patton 1947, p. 174) maxim testifies to this type of thinking. "I believe in fighting until lack of supplies forces you to stop—then digging in." To economize time, which should be a consideration of every commander who strives for efficiency of execution, such shortcuts as Patton's should not be taken. Nevertheless, Patton was well versed in manipulation of time. His approach was just so direct and no-nonsense that he occasionally omitted the finer details. In his eagerness to make most of the time and push onward he overlooked the cumulative advantages of pre-planned pauses. In his memoirs he recalled a discussion with one of his corps commanders: "Eddy proposed starting his attack on Biburg on the sixth. I told him he must attack on the fourth. He complained very bitterly that I never appreciated time and space factors. I told him that, had I done so with him or any other corps commander, we would still be west of the Seine River" (Patton 1947, p. 234). As in other areas of life and warfare, Patton broke rules to perform more within the time allocated to him. He wanted to achieve as much as possible as soon as possible and discarded less important issues on his energetic drive toward his objective. "I determined to attack as soon as I could, as I felt that time was more valuable than co-ordination. In fact, it is my opinion that co-ordination is a very much-misused word and its accomplishment is difficult" (Patton 1947, p. 255). Between these lines we can read Patton's emphasis on time and temporality, but often his overbearing and energetic personality overlooked the intricacies of the manipulation of time.

Theoretical discussion about culmination points is a relatively recent turn in operational art, but the concept is as old as war. Clausewitz considered the culmination point a necessity to determine: for the attacker to avoid overextension and for the defender to identify a suitable moment to counter-attack (Sumida 2007, p. 179). Entire campaigns and wars, such as Napoleon's failed invasion of Russia in 1812, were lost when the necessity to avoid point of culmination was ignored. Archduke Charles argued that the main cause of culmination is moving too rapidly and thus draining the fighting power of the troops due to fatigue and disorder. While this is an omnipresent threat even today, the point of culmination can involuntarily also be reached through many other mistakes in planning and execution. Vego (2009, p. VII–74) defines it as

> a point in terms of time and space reached by the attacker or the defender, after which his stated objectives cannot be accomplished and continued effort to reach them will significantly increase the chances of failure or even defeat. (...) A culminating point should not be understood too literally; it is

> not a "point," but rather an area of uncertainty or nonrecognition in terms of space and time. The higher the level of war, the larger the area of uncertainty and the more difficult it is to anticipate or sense the culminating point's arrival.

Schwarzkopf in his memoirs described providing his subordinates with his commander's intent for Operation Desert Storm. "I do not want a mechanical grind-it-out operation. We must be flexible enough to capitalize on things as they occur. The idea is not to do intermediate objectives and then stop to rearm and refuel" (Schwarzkopf 1993, p. 502). The demand for flexibility is sometimes contradictory with the methodical planning process. By setting such demands, Schwarzkopf forced his subordinates to take the difficult road. Planning mechanically that once the attack has reached a certain point, the operation enters the next phase, would have been easier, and pre-set pauses would have reduced the risk of culmination.

The point of culmination must never be reached in order not to lose both time and space. Were it so simple, no properly planned battles would be lost. Since variables at play in warfare are legion, the point of culmination shifts in time with every development in the battle. A good rule for saving time is to advance rapidly for the time determined in planning as not exhaustive of the supplies and energy of the troops and then slow down to create conditions for re-acceleration. In some cases deliberately slowed action saves time in the long run because the enemy cannot force his own rhythm on operations.

Yet, once the wave of attack breaks due to overextension, a counter wave occurs. As the Soviet offensive in December of 1941 halted the progress of German spearheads the tables were immediately turned. For two long years the war in the East was characterized by spectacular offensive gains leading to just as spectacular retreats as the other side gained enough momentum to counter them. The waves of attacks and counter-attacks taught the Soviets to set the tempo for their operations in accordance with the principle of pace with variability Liddell Hart preferred. "Stalin and his high command learned to alternate each successful advance with a very deliberate pace, to keep their armies safely short of the culminating point of victory" (Luttwak 1987, p. 24). To curtail the speed and tempo of operations, if deemed necessary, is an effective means of avoiding the point of culmination. To halt one's operations means having an "operational pause."

> In a campaign or major operation, it is rare that one's combat actions can be conducted continuously. This is particularly true when facing a relatively stronger enemy. One's materiel and human resources are always limited. Periodic slackening or even the stopping of one's forces—called operational pauses—should be part of a sound campaign plan. (Vego 2009, p. IX–125)

While operational pauses create unpredictability about one's operational tempo, they must support the attainment of operational objectives. An operational pause supplements pace with variability since tempo varies and the enemy finds it harder to estimate the progress. The activity of the troops doesn't come to a halt during a pause. A pause entails that "a major part of the main forces slackens or even drastically reduces its efforts while the supporting forces intensify their pressure on the enemy's main forces" (Vego 2009, p. IX–126). At the same time the pace of supply, maintenance, and logistics increases in support of the fighting units. Operational pauses are not only useful tools of synchronizing and sequencing operations, but "their main purpose is to allow sufficient time to regenerate one's combat potential" (Ibid.).

Occasionally there comes a need to curtail the speed of both maneuver and the decision-making cycle, but this doesn't mean stumbling headfirst into the pit Alberts et al. (2000, p. 13) warned us about when they argued that it is easy to "construct situations and circumstances where 'speed of command' is irrelevant or, worse, harmful. But there are many circumstances and missions where, all things being equal, speed of command will be decisive." One shouldn't rush to endlessly accelerate speed but estimate based on the temporal pressures whether winning time by being quicker is more appropriate than spending more time in decision-making and thus saving time in efficient execution. The right answer depends on the context, the situation, and the enemy. Network-centric warfare as a Third Wave warfighting concept attempts to provide,

> an opportunity to increase speed of command when it is appropriate; it does not force us to do so when it is not. Thus, the point we can take away is the need to better understand how we can leverage speed of command in military situations and dispel the myth that speed (or any other single factor) is either a panacea or an unmitigated good. (Ibid.)

Perhaps the most important takeaway from this discussion is the notion that it would be unwise to always be the fastest or the slowest belligerent, since part of temporizing and synchronizing the speed of planning, decision-making, and execution is the ability to find the proper tempo for each activity in each occasion. It is the art of the operational commander to understand the pulse of war and use his *coup d'oeil*, or, perhaps in this case, the inner 'ear' to set the proper operational rhythm.

3.6 Rhythm of Warfare

As human beings we need rhythm and pattern in our lives. They allow us to function but in time strict adherence to them may rob us of thought, spontaneity, and reason (Stempel 2012, p. 62). Following patterns in all human activities leads to routine repetition which should be avoided in operational art. Rhythm of warfare is a seldom discussed but important concept. In the agrarian age having rhythm saved time, since soldiers could move as a unified army because their footsteps followed the same rhythm. The Spartans marched slowly to the rhythm of flutes not to break their ranks (Thucydides 1971, p. 351). So did the Romans, but this practice was forgotten for a long time (von Bülow 2013, 238). De Saxe (1987, p. 202–203) wrote that "it is a comedy to see even a battalion commence movement! [...] Have them march in cadence. There is the whole secret, and it is the military step of the Roman. That is why these musical marches were instituted, and that is why one beats the drum." The rhythm of the drum enabled synchronized movement, which was important in keeping formations together, but also when armies marched to battlefield. Since they marched in cadence, one could calculate how long a march would last and it became possible to time the arrival of separate formations to the battlefield like Napoleon did.

In our contemporary conjunction with its fetish of accelerating the pace of operations and reducing response times, the idea of natural rhythm is overlooked. Freytag-Loringhoven emphasized that "the leader must be sensitive to the pulse of the battle" and discussed the view General Kuropatkin as his chief of staff held of Skobeleff. Allegedly he had a skill "to feel the pulse of the battle accurately—now rapid, now slow—and how Skobeleff would shape his own conduct in action accordingly, sometimes holding back, checking his own impulses by the power of his will, sometimes giving them free rein" (von Freytag-Loringhoven 1991, p. 208). The expression "pulse of the battle" refers to an inner rhythm directing the course of the battle. The rhythm of a battle or an operation is not constant, but like a pulse, speeds up with exertion and slows with relaxation.

Treating the rhythm of the operation as a pulse of some magnificent beast provides insight into its regulation. The commander adjusts his own pace of action to correspond to the pulse of the operation when the rhythm takes over. Yet he has a possibility to re-set the pulse with his activity. This is what planning of operations essentially is. When the commander orders the execution of his plan, he sets the rhythm for his operation, but he cannot keep the pulse constant beyond the period of time when battle proceeds according to plan. Exertion in faster maneuver elevates the pulse and over-exertion causes it to stop. Meanwhile the enemy constantly seeks to alter this pulse to favor his operational purposes. Once the planned and actual courses of the battle no longer are in unison, the pulse of combat takes over and starts to dictate its own rhythm and influence the rhythm of operations alike.

This is another *kairos*-time for intervention by the operational commander. He still holds the reins, and by judging and choosing his action properly and initiating it at the right time, he can speed or slow down the rhythm. He cannot, however, allow his inner need for speed to dictate his actions. He must set his own rhythm to follow the pulse of the operation and seek a proper moment to shock the battle to a new, artificial, rhythm or initiate a new battle to reset the rhythm of the entire operation. As an example suffices the moment when defense switches into offense. A slow rhythm is replaced with a fast one. This transition, however, cannot be made unless the commander has abided to a slower rhythm to start with. Every event has the potential to alter the natural rhythm of battle and the commander must have his finger on the pulse to deduce the alterations these events constantly cause. This requires a profound understanding of the tempo of warfare. Battle, operation, and war alike adhere to rhythms of their own.

> War is a pulsation of violence, variable in strength and therefore variable in the speed with which it explodes and discharges its energy. War moves on its goal with varying speeds, but it always lasts long enough for influence to be exerted on the goal and for its own course to be changed in one way or another. (von Clausewitz 1989, p. 87)

War had a rhythm for Clausewitz as well, hence the "pulsation." Intensity and speed of warfare vary, while war progresses toward its final resolution.

Sometimes warfare is slow, sometimes rapid, sometimes time stands still, at other times it races and runs out. Even when there seems to be no time for anything, leadership and influence need to be exerted to change the situation. These moments when time flies are when crucial decisions are made in operational art. When time is stagnant and there is an abundance of it at the commander's disposal, the enemy commander enjoys the same advantage and the possibility to create something unexpected diminishes. When time is scarce, a true operational artist exerts his superiority over a lesser general.

Temporizing maneuver requires more than moving faster or slower in response to operational needs. Proper maneuver both prior to battle and during it adheres to a rhythm. "The movements of armies resolve themselves into a constant separation and reunion. For both the right moment must be chosen. If the forces are concentrated too soon, it will be necessary either to disperse again, or to march in close formation over a narrow space and upon few roads" (von der Goltz 1906, p. 174). The rhythmic pulsation in warfare is both temporal and spatial and occurs within an operation as well.

> Temporal distribution (also known as preemption) is the temporal converse of concentration. Preemption sacrifices combat power to achieve a temporal advantage, with a view to attacking an unready enemy. Concentration sacrifices time in order to garner combat power, with a view to attacking a ready enemy (Leonhard 1998, p. 257).

By choosing the right time to disperse and concentrate his force, the commander maximizes the effect of his troops. Proper temporizing turns execution of operational art into a dance of separation and reunion of forces. If troops are separated, when they should attack, their action is not unified and their potential combat value is wasted. If they are massed together under enemy fire, the damage received grows in proportion to the level of concentration. The art of concentration of forces in space and time

> consists not simply of concentrating our own forces but by causing the enemy to disperse his. To bring this about, it will normally be necessary for us to disperse our own forces so as to confuse the enemy and draw him away from our real objective. Thus concentration actually consists of dispersion, whereas dispersion consists of concentration, victory going to him who,

> retaining control and avoiding confusion switches rapidly from one to the other. (van Creveld 1991, p. 121)

The times and places when and where units should converge and disperse have to be identified during the planning of the operation. The purpose in operational art is to use variability and not to choose one or the other but to employ both in turn and use this pulsation to create discord in the rhythm of the enemy. The idea is old but seldom fully exploited.

> A proper and effective balance between concentration and its two converses, spatial and temporal distribution, leads to success. An overreliance upon either extreme will just as certainly lead to disaster, because the enemy will react to diminish the effects of either. The commander in war should structure his plan to alternate between distribution and concentration. (Leonhard 1998, p. 257)

Clausewitz saw warfare as a spasmodic process in which operations and battles are interrupted by periods of non-action when both sides observe each other on the defensive. One side is usually more motivated than the other and chooses to use the dominating element of offense, which will result in a clash and continuation of action (von Clausewitz 1989, p. 219). The subjective notion of time is different during the spasm and the relaxation. When action increases, time flies and during the period of non-action, there seems to be an abundance of time. Neither ceaseless action, nor absence of activity is a suitable state to uphold in an operation. There must always be some activity and even more importantly the enemy must be kept short of time.

> The enemy must be forced to dance to your tune all the time. This means that the commander must foresee the battle. He must decide in his own mind, and before the battle begins, how he wants operations to develop; he must then use the military effort at his disposal to force the battle to swing the way he wants. My own military doctrine was based on unbalancing the enemy by manoeuvre while keeping well balanced myself. (Montgomery 2000, p. 17)

According to Miyamoto Musashi, one cannot control the timing of strategy without relentless practice. To have this ability to fully control actions releases strategy from insecurity. For Musashi the entire life of the warrior and strategist expresses timing. He grows, matures, and withers. Everything is connected to temporality and this needs to be recognized by the strategist. He needs to discern proper and improper timing and among the fast and slow tempos of great and small events alike choose the most fitting time for action. This requires knowing the timings on the background of unfolding events. Timing is the heart and soul of the art of war and by knowledge of "emptiness" the strategist will emerge victorious from any battle. It means knowing the timing your enemy uses and utilizing a timing of your own that the enemy cannot expect (Musashi 1995, p. 58).

For Musashi the timing of warfare was always joined to a rhythm. Events on a battlefield have a rhythm of their own and one's actions need to be synchronized with the rhythmic superstructure of operations (Cleary 2005, p. 38). One needs to, however, occasionally reset the rhythm and this is where superior timing exhibits itself. The rhythm of the enemy has to be exploited to cause confusion among his troops when his situation is unstable (Musashi 1995, p. 97). A discord has to be introduced into his rhythm while maintaining one's own. If the battle has reached a stasis, one has to cease his effort, apply a new perspective to the situation, and pick up a new rhythm for the continuation of the operation. Something unexpected is required to bring the enemy again to the brink of disequilibrium and force him to follow one's own rhythm and timing (Musashi 1995, p. 89).

Time seems to have a disturbingly dual nature in operational art. There is a need to seize the moment, when it arrives, but one can occasionally wait for it for a long time. There is no time to be wasted when a decision is required, but no inherent reason to save time before it. There will at some point be increased activity, but awaiting that, lackadaisical passivity of *kronos*-time is the prevailing condition. There are no fixed rules for time, just guidelines how it should be treated. The same applies to speed and haste. "It is the nature of the army to stress speed; to take advantage of the enemy's absence; to travel unanticipated roads; and to attack when they are not alert" (Sun-tzu 1993, p. 179). Simultaneously there should not be haste since, "to race forward day and night without encamping, covering two days normal distance at a time, marching forward a hundred li to contend for gain, the Three Armies' generals will be captured. The strong will be first to arrive, while the exhausted will follow" (Sun-tzu 1993, p. 169). Winning time may have a cost. Rapid movement should be strived for without eroding the capabilities of the army. This is evident in the early European masters of war too. For example, Frederick strove to deliver his attack on an unprepared enemy through precision of movement and not by hurrying (Colby 1943, p. 80).

Sun-Tzu (1993, p. 183) advised that "if the enemy opens the door, you must race in." Since he also advocated knowing oneself and the enemy, he doesn't advise rushing in to fall prey to a feint or ruse. Rush-in must occur when conditions for it have ripened (Yuen 2014, p. 91). Wu-tzu (1993, p. 217) offered additional advice: "courage is but one of a general's many characteristics for the courageous will rashly join battle with the enemy. To rashly join battle with an enemy not knowing the advantages and disadvantages is not acceptable." It would be folly to attack unthinkingly. A moment may look auspicious, but it can be a stratagem of the enemy. Thus, one must "ascertain the enemy's voids and strengths and then race [to take advantage of] his weak points" (Wu-tzu 1993, p. 213). The duality of yin and yang appears in strategic and operational thought. It is the universal logic of war and strategy (Yuen 2014, p. 28–29). To describe how the army should be used, following this duality, "its speed is like the wind, its slowness like a forest; its invasion and plundering like a fire; unmoving, it is like the mountains. It is as difficult to know as the darkness; in movement it is like thunder" (Sun-tzu 1993, p. 169).

A good operational commander must combine in his rhythm haste with prudence and speed with stability into rapid, well thought-out maneuver. One has to win time to catch the enemy unprepared, but prepare oneself thoroughly. Time can be squandered as well as stolen from the enemy, but at the heart of everything is painstaking preparation to identify the advantageous moment and to seize immediate initiative. For Musashi (1995, p. 80) finding the auspicious timing is not concerned with the maxim of striking first. A suitable timing and proper rhythm is also to attack concurrently with the enemy, only harder and using the distraction caused by your attack to emerge victorious. If the enemy is able to beat your timing with his attack, you need to retaliate. If the attack of the enemy is rapid, your response needs to be strong and calm, even slovenly. If the enemy uses a slow attack, then your response needs to be swift and powerful.

Every commander is naturally poised to attempt to seize the initiative from his enemy. If that head start cannot be achieved, one should not aim to respond with the same speed, since the element of surprise belongs to the one who initiated action. To respond timely with grace, with no hurry, but careful preparation, will enable one to regain the initiative. It is a conscious decision not to subject oneself to the timing of the enemy, but to act with a level head in unison with the rhythm most suitable to oneself.

Speed is identified with spontaneity, but a precise response is different than an automatic one (Cleary 2005, p. 58–60). The lower the level of warfare we discuss, the smaller is the opportunity to use one's own rhythm since tactically a quick response may be a necessity. But in operational art the response to enemy action may win time by avoiding undue haste. If the speed is too high and the actions are not properly synchronized one will break his own rhythm. For Musashi speed doesn't play a role in strategy, because an omnipresent part of the idea of speed is that things look fast or slow depending on their rhythm. The master of strategy never looks fast, but has exquisite timing. If one acts too slowly, one is not able to exploit the confusion of the enemy to decide the battle. The master of strategy acts at the right moment, using immaculate timing and appropriate speed without haste in accordance to his own rhythm that emerges from his thorough knowledge of the way of strategy (Musashi 1995, p. 99). There should be continuous interaction between one's own army and that of the enemy. Only then the right moments to cause discord in the rhythm of the enemy can be exploited. One should

> endanger them to observe their fears. Be tranquil to observe if they become lax. Move to observe if they have doubts. Mount a surprise attack and observe their discipline. Mount a sudden strike on their doubts. Attack their haste. Force them to constrict their deployment. Launch a sudden strike against their order. Take advantage of [their failure] to avoid harm. Obstruct their strategy. Seize their thoughts. Capitalize on their fears. (Ssu-ma 1993, p. 142)

In First Wave warfare of the agrarian Orient such interaction between armies was easier to carry out during the operation than in indust-real times of greater mobility. The most profound guideline for setting the tempo is Ssu-ma's order to "attack their haste." When the enemy is pressed for time and forced to hurry, then is the suitable time for attack. More importantly, during the times he is forced to perform in a hurry, he is open for a surprise attack in turn. This should lead the operational artists of today to consider if the most prudent way to wage war is to constantly increase speed to allegedly press the enemy for time. At least on occasion choosing a rhythm and timing that comes naturally to the commander and his troops may be more fruitful. This remains true especially during the Third Wave where the idea of time in operations is more flexible.

3.7 Synchronizing Operations

> The art of orchestration requires many skills not traditionally found or emphasized in the military: diplomacy, patience, consensus building, and imagination. Of these, the last is the hardest to cultivate. Orchestration requires imagination, because, just as the conductor of a symphony hears the music in his head—hears what it ought to sound like—and waves his baton accordingly, so the military commander must be able to imagine what the end state of his forces' efforts should be. (Leonhard 1998, p. 203)

To make the best of rhythm of operations the commander has to synchronize the use of assets at his disposal. Just creating a timetable of activities would not constitute the level of orchestration Leonhard wrote about. This requires imagination and inspiration to go beyond planning an operation to producing operational art. A challenge for armed forces of all times is to combine thorough planning of operations with effective execution while adding the artistic touch. The combination of a detailed plan and a chaotic execution has been witnessed often enough. Sherman (1890, p. 209) wrote that the Battle of Bull Run "was one of the best-planned battles of the war, but one of the worst-fought." All too often during the American Civil War the planners did not consider realities on the ground like terrain, logistics, capabilities, and, most importantly, time (Stoker 2010, p. 9). Problems emerged when real life refused to conform to the plan. The plan must have potential for further development according to the emerging situations in the course of the operation.

Operational planning is a time-consuming process, but if conducted properly, a huge timesaver as well. As Jomini (2007, p. 31) wrote, "concert in action makes strength; order produces this concert, and discipline insures order; and without discipline and order no success is possible." To enable quick victories, operational plans need to be painstakingly crafted. Therefore, one cannot artificially attempt to produce plans faster. One can, however, increase the efficiency and streamline the planning process to ensure that no time is lost. Nevertheless, the more economical, efficient, and streamlined any organization is, the greater is its vulnerability (van Creveld 1991, p. 121). Planning time can be reduced

> if the planning is conducted concurrently rather than sequentially. Also, the simpler the planning process, the more speed is enhanced. In addition, the more experienced the operational commanders and planners, the faster they will be at improvising and preparing operation plans. (Vego 2009, p. III–25)

The traditional mode of planning meant that one echelon of command prepared its plan and then distributed this product de facto ordering the subordinates to initiate their own planning cycles. After these plans are finalized, they are again given as orders to the next echelon of command and so on. The process is simply too long to be applied when time-pressure affects every link in the chain of command. Time is saved by making the process concurrent, that is, getting subordinate commands to participate in higher level planning while simultaneously creating their own plans. By involving subordinates in the planning phases where their tasks are drafted, the commander's intent becomes clearer. Being given a ready-made order as a basis of one's own plan and finalizing it in turn before assigning tasks to subordinates is the simplest procedure, but the simplicity lost in concurrent planning is amply rewarded by gained time. To win time in planning requires a balance between simplicity and effectiveness.

We have some tools, techniques, and methods pertinent to the Third Wave in our doctrines, but we still tend to view them as separate elements to be coordinated or attempt to use them in traditional ways. This incompatibility also concerns the way we conceive of operations. As Alberts et al. (2000, p. 70) describe the treatment of operational art,

> Our current approach to developing a military campaign plan is predicated upon a fairly well understood set of relationships among events that take time to unfold. Thus, the plan can be decomposed into a series of steps, each one building in a linear fashion on the preceding steps. Our ability to deal with something as complex as a military campaign depends upon our ability to break it down into these manageable pieces. We can do so because of our ability to separate events in time and space. Organizationally, we deal at three levels—the strategic, operational, and tactical. Geographically, we deal with sectors or theaters. Functionally, we usually deal with specific jobs or tasks in a sequential manner (e.g., first we do suppression of enemy air defenses and achieve air superiority, then we attack other targets). The battlespace is thus segmented, and we can deal with smaller isolated problems, tasks, or battles. The nature of Information Age Warfare makes it more and more difficult to operate in this reductivist fashion. Technology has compressed the space and time continuum, and political realities have collapsed the clear separations among the strategic, operational, and tactical levels by introducing more dynamic rules of engagement.

Once again, the reader must benevolently allow such a long quotation. It is included to allow the writers to express their idea of simultaneous actions of the information age in relation to sequential actions. There are two types of strategies, one is sequential and proceeds by discernible steps. The other is the cumulative and in it the accumulation of actions may lead to critical mass. These two can be used simultaneously as they don't contradict each other (Wylie 2014, p. 24–25). In operational art sequencing is a practical necessity. Sequential nature of operations was the result of industreal willingness to divide and subdivide everything mechanistically into smaller units. This enabled commanders to comprehend and control operations. Different pieces and phases connect and the causal chain of action and reaction becomes manageable. Operations have grown into complex entities that need to be broken down into component parts the planners can first tackle individually and then combine into an operation.

> Planners use major phases to break their tasks into more manageable parts. The major phases in the employment of one's combat forces at the operational level are mobilization, predeployment, deployment, combat employment, posthostilities, redeployment, demobilization, and reconstitution. The initial campaign would most likely encompass all of the major phases. The separation between the successive major phases is usually blurred. (Vego 2009, p. VI–3)

Artificial division of operations into separate phases supports planning. During the execution of the plan, phases blend together, but the commander can follow closely how the operation proceeds and whether intermediate objectives have been reached and order the switch from one phase to another. Within a phase of operations simultaneity is desirable.

Sequencing is an important part of synchronization. The entire development trajectory of RMA, EBO, NCW, and military transformation has followed the idea of waging war by exploiting new technologies to create tactical effects which in turn should be carefully sequenced, so that the combined effect would have operational impact (Strachan 2013, p. 214). Warden wanted a strategy to shorten the war and he advocated conducting operationally a "parallel attack" which meant basically operations synchronized in time and space. In his words, once the enemy system is understood, crucial systematic vulnerabilities have been found, and the desired effect is decided on, "parallel attack will normally be the preferred approach, unless there is some cogent reason to prolong the war" (Warden III 1995, p. 19). A state has a limited number of vital strategic targets that have no backups or are difficult to repair. If a significant percentage of them can be attacked simultaneously, that is, in parallel, the damage reaches the breaking-point of the "system." Parallel attack aims to concentrate violence in time so that the shock effect increases.

> The enemy can alleviate the effects of serial attack by dispersal over time, by increasing the defenses of targets that are likely to be attacked, by concentrating his resources to repair damage to single targets, and by conducting counteroffensives. Parallel attack deprives him of the ability to respond effectively, and the greater the percentage of targets hit in a single blow, the more nearly impossible his response. (Warden III 1995, p. 19–20)

Hitting tens or hundreds of parallel targets simultaneously requires a vast array of capabilities and practically demands a quantitative and qualitative superiority such as what a superpower can amass against a small state. In the past and even present, serial attacks have been the traditional way to conduct operations because

> parallel attack has not been possible on any appreciable scale in the past because a commander had to concentrate his forces in order to prevail against a single vulnerable part of the enemy's forces. If he prevailed, he could reconcentrate and move on to attack another point in the enemy's defenses. The process of concentrating and reconcentrating was normally lengthy and one that the enemy worked hard to foil. This process, better understood when labeled "serial warfare," permitted maneuver and countermaneuver, attack and counterattack, and movement and pause. (…). All of our thinking on war is based on serial effects, on ebb and flow. The capability to execute parallel war, however, makes that thinking obsolete. (Ibid.)

Clausewitz (1989, p. 206) gave guidance into timing one's operations by arguing that "in the tactical realm force can be used successively, while strategy knows only the simultaneous use of force." In operational art of the Third Wave one should attempt to combine the successive with the simultaneous. Synchronized parallel attack is an ideal that has been the objective of the art of war ever since first tactical formations appeared on the battlefield. Synchronization requires all elements, be they men or machines, to act in concert, but theories don't yet correspond with reality because this level of synchronization often remains unattainable. Transforming the concepts of operational art from sequential to self-synchronizing actions would theoretically enable moving from

> a series of static events to a more continuous one by greatly increasing the operating tempo of events. The result is the need for greater integration between the heretofore separate planning and execution processes, requiring more timely interactions between the two, and portents an ultimate merging of these two processes into a seamless form of command and control. (Alberts et al. 2000, p. 74)

While they are right in their claim that planning and execution are currently considered as two different sequences of operational art, our practices tend to let both of them operate in 'loops' of their own. In other words, executing one phase of an operation and planning the next are two different process-cycles that simultaneously revolve on their own. Instead of "merging" these two into "a seamless form of command and control" there exists a need to synchronize them so that output of one acts an input of another. The principle of simultaneity should not concern only simultaneous effects. There should be simultaneous activity on all the levels of warfare to provide input for other levels. This requires synchronization so that in the operational picture activity is unceasing and no time is wasted by breaking it into consequent phases even if this often reflects reality on the tactical level.

> Viewed over time, the activities at the different echelons take place sequentially, with one level executing the existing plan while another is developing the new plan. This process has evolved to the point where planning and execution are distinct activities. Efforts to speed up the process so that more responsive plans can be developed are fast approaching the laws of diminishing returns. (Alberts et al. 2000, p. 75)

Cycles of both processes have accelerated beyond their natural limits and they still lack synchronization. On occasion the speed of either or both has to be slowed unless there is a pressing operational need. The idea that the cycles of planning and execution are intrinsically interconnected like clockwork is a remnant of indust-reality. The self-synchronization NCW strives for must allow these cycles to regulate their own speeds and if they fail, the commander's intent and orders will provide the rhythm for both cycles.

Self-synchronization supposedly is "a mode of interaction between two or more entities" (Alberts et al. 2000, p. 175). But if due to omnipresent friction in war, the command structure is unable to self-synchronize itself optimally, the commander resets the rhythm. Warden (2000, p. 109–127) wrote a whole chapter dedicated to "orchestration of war". To use allegories of rhythm and an orchestra, we see in sequential or serial operational activities not a harmonious symphony but rather different compositions and instruments played one after the other. Third Wave enthusiasts claim that NCW enables synchronization of activities and empowers the network to self-synchronize. This sounds utopian, since even a symphony orchestra requires a conductor to lead it and in a military 'symphony' this is the task of the commander. He makes the instruments play in harmony, complementing each other to maximize their effects.

> Having selected an objective, the composer decides how best to reach that objective. Should it be a piano concerto, a violin concerto, or a flute concerto? (…) The composer, and later, the director, has the task of orchestrating—not subordinating nor integrating—his instruments so that each can do its job—whether that be as the key force or the supporting force. In the process, he does not try to make one instrument sound like another, or do another's job; rather he uses each to do what it is naturally constituted to do and what only is capable of doing. (Warden 2000, p. 123–124)

Therefore, the timing of each instrument of war and their characteristics have to be synchronized to complement each other. Simpkin (1985, p. 94) noted that "music communicates wholly at the aesthetic level and depends entirely on its dynamics to do so." If dynamics of time and space within the battlespace conjoin, the operational artist needs to retain control of the increasingly fluid and dynamic situations. Operations are conducted at least partially in sequential manner, since every commander tasks his subordinates with specific tactical tasks so that the combined effects will fulfill his operational objectives. Often it is not even feasible to try to strive for simultaneity. Leonhard (1991, p. 177–178) wrote that modern combat is

> a contest for time. But the need for the simultaneous executions of two separate actions must not be assumed in every case. Often the precise arrangement of attacks in sequence rather than simultaneously produces that sought-after effect known as combat multiplication. (…) Further, in emphasizing simultaneous events, it robs the military planner of appreciation for the powerful potential of sequential events.

In the strategic picture the spatial and temporal sequences combine harmoniously. On operational level the task is to combine the isolated activities in time and space to enable creation of this harmony. In addition, the ones responsible for operations and adapting the plan must understand the dynamic nature of the different actions or sequences, so that the interrelations between components are clear. In this way once chaos of discord enters the harmony, as friction of war is bound to cause to happen, the commander is poised to understand the nature of the discord, locate where it occurs, and how it affects the rhythm of the whole. Then he can synchronize the source of discord with the rhythm of the battle again. While discussing music in relation to the rhythm of maneuver in battlespace may seem far-fetched, it might again be beneficial to quote Simpkin who wondered how "it never ceases to strike me that the modern world's three most successful practitioners of manoeuvre theory, the Germans, the Russians and the Jews, are also the most musical peoples of our civilization." (Simpkin 1985, p. 94)

3.8 Asymmetric Warfare as Variable Operational Art Temporality

> Whenever possible, the operational commander should use friendly forces asymmetrically and plan for operational/strategic deception. To preserve versatility and variability of decisions, the operational commander should never act according to conventional views and preconceived notions. The art of warfare rests on the freest application of its fundamentals under constantly changing conditions. (Vego 2009, p. X–55)

The idea of variability of pace can be extended to other areas of operational art as Vego argued. The primary goal of variable speed is to deny the enemy the ability to predict the maneuvers correctly and synchronize his responses. When an entire operation is planned asymmetrically many other factors become unpredictable as well. Breaking conventions of military dogma through versatility and variability has always been the key to asymmetry and freedom of application of the principles of war. Viet Cong's "armed forces always act unpredictably in fighting the enemy, unpredictably in their direction of attack, their target, time, use of force, scope of attack, manner of attack, and so forth" (Giap 1970, p. 82). Unpredictability can be an asset, if one manages to remain constantly original and surprising in one's operational art—as some great captains of history did. This variability is evident when Delbrück (1990a, p. 231) described how Alexander "combined the various arms in a different way each time, according to different circumstances, for the strongest possible total effect."

The discussion concerning asymmetric warfare has for years focused on the use or terrorism, weapons of mass destruction, dirty bombs, and similar tactics adopted by illegitimate fighters. Yet, asymmetricity should be the objective of every operational plan of conventional forces as well. This is because "the asymmetric application of available combat potential offers the best opportunity to achieve quick and decisive results, even when the stronger side uses it" (Vego 2009, p. IX–113). The purpose of asymmetric warfare is to be strong where the enemy is weak and target those weaknesses with all available means that preferably differ from the ones the enemy employs. At least since the middle of the nineteenth century U.S. operational art has focused on offensive operations aimed at defeating the enemy's main forces, pitting strength against strength (Echevarria 2011, p. 139). It was a bloody method, but it serviced the two World Wars. Problems emerged in Vietnam since there was no tangible enemy to hit. New methods had to be learned but they could as well have been adopted from history.

From the operational perspective the ancient Indian Kautilya (book VII, ch. 6, p. 15) saw three different ways of engaging the enemy. "Open battle, treacherous battle, and silent battle (i.e. killing an enemy by employing spies when there is no talk of battle at all), are the three forms of battle." These three forms of fighting occur in a reversed order from the one here. Silent battle, or, not to mince words, political murders and assassinations, comprise the initial stage of war. They are used before warfare proper and continue from the decision to commence hostilities at least until the actual combat begins, when assassinations become more difficult. The time to use treacherous battle is when armies march out to meet on the battlefield. "The beginning of an attack is the time for treacherous fights" (Kautilya, book X, ch. 3, p. 3–4). The three types of battle Kautilya described remain relevant because silent and treacherous battle belong to asymmetric warfare and increase the fighting power of those who seek to avoid open battle. In contemporary context

> asymmetrical action means employing one's force against a dissimilar hostile force, as, for example, naval versus air or land forces. The objective then is to use the strengths of one's forces against enemy weaknesses. Asymmetrical

> actions usually generate disproportionate outcomes and thus have the potential to produce a quick and decisive victory with minimal losses. (Vego 2009, p. IX–149)

A tank is wasted when it has to fight a tank. Pitting infantry against infantry leads to huge casualties. Simpkin (1985, p. 174) argued that whenever two similar but unequal machines of organisms fight each other directly, the law of probability dictates that the stronger, or the more numerous, will almost inevitably win especially if they follow the same modus operandi. Placing dissimilar opponents against each other changes the rules and creates a different kind of contest which even the weaker can win. Asymmetric warfare provides a way to avoid fighting at a time and place of the enemy's choosing where he can utilize his strengths. Against a powerful state a small state can have a comparative strategic or operational advantage only in low-intensity and low-tech methods of war (Handel 2001, p. xxii). There is a long tradition of asymmetric warfare is China (Scobell 2011, p. 197). The way of asymmetric warfare and its benefits can be summarized thus:

> Do not oppose the dense with the dense; do not oppose the dispersed with the dispersed; do not oppose the full with the full; do not oppose the vacuous with the vacuous; do not oppose the urgent with the urgent; do not oppose the slow with the slow; do not oppose the numerous with the numerous; do not oppose the few with the few; do not oppose the rested with the rested; do not oppose the weary with the weary. (Sun Pin 1995, p. 225)

The avoidance of having like fight like is the bottom line of asymmetric warfare today. A good WW II example of a tactical level failure to do this was the battle of Kursk where massive tank armies fought each other almost to the devastation of both. Asymmetry could have made the battle less destructive. Symmetry and asymmetry exist on all levels of warfare. In operational art it is about choosing the forces to pit against each other, deciding the methods employed, and choosing a rhythm for operations that is harmonious with the objective. On the strategic level, by choosing,

> a symmetrical strategy, a country or group of countries tries to match its superior strength against that of a similar enemy that is unable to adequately respond in kind. An asymmetrical strategy aims to counter the enemy's

> strengths and to accentuate and exploit his weaknesses and vulnerabilities by applying nonconventional means. (Vego 2009, p. I–41)

On occasion the idea of asymmetry is discussed through different terminology. Leonhard approached the dilemma through the dualism of "objective" and "subjective" warfare. For him subjective conflict is the traditional way of waging war while

> Objective conflict, on the other hand, occurs when an opponent applies combat power against an unlike and vulnerable aspect of the enemy. (…). Objective conflict pits strength against weakness and vulnerability, or, in a more general sense, it applies energy to something other than competition with a counterpart. At the technical/tactical level of war, it means fighting against an "unlike" system. (Leonhard 1998, p. 229)

For the purposes of clarity, we will discuss symmetry and asymmetry as possible choices of tactics or operational art. But tactics resonate on other levels and thus the "choice of tactics is in fact strategy. It is strategy which decides the form in which the conflict is to be waged, whether it is to be offensive or defensive, whether it will use force or subversion, whether force is to be used directly or indirectly" (Beaufre 1965, p. 48). Beaufre in his texts discarded the operational level where the majority of tangible practices of asymmetry are located. To illustrate this, an excursion into one asymmetric operational choice is needed.

Guerrilla warfare in its different forms is the most widespread asymmetric method of fighting. It abides to its own rules of temporalities and has an utterly different perspective on how time should be managed than regular warfare. While regular armies in their operational art seek decisive battles to finish the war quickly, guerrillas are often the ultimate attritionists. As Lawrence (1997, p. 95) wrote about the Arab War and his irregular troops, "Our aim was to seek the enemy's weakest material link and bear only on that till time made their whole length fall." Operationally guerrillas seek to prolong the war as long as is necessary for the enemy material base and ultimately the will to fight to collapse. These opposed perspectives and objectives lead to differences in thought and often the guerrillas are operationally better than regular armies in thinking outside the box. Liddell Hart described this relationship so eloquently that he deserves a longer quotation. He claims that the difference between regular and guerrilla forces is

> in the rarity of inspired and creative leadership in the organized armies of history. Generals have been legion; artists of war few. Many more, and infinitely more military genius can be traced in the scantier records of irregular and guerrilla forces. The obvious explanation is that the natural gifts of these leaders who have emerged straight from the womb of conflict, instead of a professional incubation chamber, have not been cramped or warped by convention. To acknowledge the genius of many of these irregular leaders does not imply that they would have been equally successful if placed in command of a more highly organized type of force, at any rate until they had gained experience of its organization and different application. (Liddell Hart 1937, p. 145)

While guerrillas or at least the type of asymmetric methods they tend to favor have been evidenced in most of the conflicts of history, they have left relatively little for students of the art of war to peruse. From the mass of popular uprisings in which guerrilla-type tactics were abundant only a few clear and concise theories emerged. Fanon's ideas from Algeria are written down in novels, some others stuck to writing diaries, but Guevara, Mao, Giap, and Ho Chi Minh left their instructions to future insurrectionists. In sprit and context Mao's work is aligned to Clausewitz in many senses (Handel 2001, p. xxiv).

> Select the tactic of seeming to come from the east and attacking from the west; avoid the solid, attack the hollow; attack; withdraw; deliver a lighting blow, seek a lightning decision. When guerrillas engage a stronger enemy, they withdraw when he advances, harass him when he stops; strike him when he is weary; pursue him when he withdraws. In guerrilla strategy, the enemy's rear, flanks, and other vulnerable spots are his vital points, and there he must be harassed, attacked, dispersed, exhausted and annihilated. (Mao 2000, p. 46)

Mao illustrates how advantage can be derived from the asymmetry and time factors. The emphasis on "deliver a lightning blow, seek a lighting decision" shows that a guerrilla operation should be a single flash of fighting, causing maximum damage and ending almost as soon as it struck. Timing of the attack should ensure the unpreparedness of the enemy. And the guerrilla must withdraw before the enemy can retaliate. Successive attacks in different loci are unceasing and there is an element of constancy involved. The enemy needs to be "harassed, attacked, dispersed, exhausted and annihilated" at all times.

Picking and choosing the advantageous time depends on when the enemy is most disadvantaged. If his troops are tired, dispersed, or resting, then the guerrilla attacks with strongest possible force. Whenever the enemy is strong and ready for battle, the guerrilla only irritates him and erodes his morale by quick and elusive attacks followed by immediate retreat. For irregular troops, "dispersal was strength" (Lawrence 1997, p. 95). These attacks work to weaken the enemy, and only when the enemy is weak enough, the guerrilla will strike with concentrated force to seek a decisive battle of annihilation. In the words of Ho Chi Minh, (2008, p. 27) "look far ahead and ponder deeply. Be resolute: attack and attack incessantly." Small operations slowly erode the enemy. When the timing looks promising the guerrilla will hit with amassed force, but remains always ready to withdraw. Thus judging the opportune moment for a particular type of action is the foundation of effective guerrilla operational art. Such an asymmetric situation leads to what Tukhachevsky called "the battle of the bugs". There is no decisive battle but

> gradually the invading army loses its vitality; more and more it feels the need to bring the occupation to an end. For the professional army this is indeed one of the most difficult situations. But it remains to be seen whether the concept of a 'small war', a battle of the bugs, can really be applied to the struggle of two opposing organised forces, each capable of covering its rear. Surely it cannot! (cited in Simpkin 1987, p. 87)

Just because guerrilla warfare is unorganized and both its structure and discipline differ from conventional army procedures, it often is distasteful for professional soldiers. Perhaps the difficulty to counter it with conventional tactics plays a role too. As Ludendorff (1919, p. 25) wrote about WW I: he had expected a humanistic and chivalrous war, but ended up disappointed and disgusted by the asymmetric warfare he experienced. Leonhard (1998, p. 230) claimed that when traditional military mind "catches a glimmer of objective conflict, he is quick to set it aside as a nonmilitary—even distasteful—phenomenon." Nevertheless, when Germany was militarily weak in 1930s Ludendorff turned around to strongly support *Volkskrieg* in which civilians become fighters (Corum 1992, p. 64). Asymmetric methods are effective and thus sometimes necessary and even popular.

Asymmetric warfare is distasteful to some of its practitioners as well since it may not be their preferred method of fighting. The choice might not be a cultural preference so much as a political necessity (Coker 2010, p. 229). As Paret (2015, p. 108) wrote, "Tactics often give voice to a social or political reality; at other times they adjust to the most varied conditions." Culture of the fighters has a huge impact. Lawrence (1997, p. 301) wrote how rare it was in the Arab War for a professional soldier to have "married war and rebellion in himself" as he wanted every British officer to do. A professional soldier prefers to use the methods he has been taught and as an operational artist deploys his troops accordingly. Resorting to guerrilla warfare is often a choice enforced on the weaker side. When the weak has eroded the power of his enemy while strengthening himself he tends to adopt traditional methods. "The war of liberation of the Vietnamese people was a long and vast guerrilla war proceeding form simple to complex then to mobile war in the last years of the resistance" (Giap 1962, p. 48).

Ernesto 'Che' Guevara contrasted the "rigidity of classical warfare" to guerrilla warfare where the guerrilla "invents his tactics for each moment of battle and constantly surprises his enemy" (Guevara 1962, p. 17). The same ideal of reinventing tactics and operational art to suit the requirements of a given moment is a valuable skill for anyone in command of conventional forces as well. Asymmetry is an asset if utilized properly. If it is a 'dirty' way of fighting, does it matter when chivalric warfare is a thing of the past anyway? As Ho Chi Minh asked, "Isn't manure dirty? But if it's good for the rice plants, will you refuse to use it?" (Cited in Giap 1968, p. 109). The idea of guerrilla warfare is to stand opposed to the practices of the conventional armed forces and use their weaknesses. Even in operations and tactics the guerrilla does exactly the opposite as his enemy. "The enemy advances, we retreat; the enemy camps, we harass; the enemy tires, we attack; the enemy retreats, we pursue" (Mao 1963, p. 70). Common sense dictates that asymmetry should be used more extensively in operational art to counter enemy stratagems.

In asymmetric warfare forces should not be employed as a whole and preferably not at all against a sizable conventional enemy force. They should not be used, as Clausewitz (1989, p. 481) put it, "to pulverize the core but to nibble at the shell and around the edges. They are meant to operate in areas just outside the theater of war—where the invader will not appear in strength—in order to deny him these areas altogether." In more recent times Giap framed the same thing in other words, emphasizing the need to reduce the enemy strength bit by bit and pave the way for an attack with conventional forces. "We must erode, annihilate, disperse, and [harass] the enemy everywhere, creating conditions for mobile forces to launch concentrated blows to annihilate him wherever his gaps are exposed" (Giap 1970, p. 34). Guerrilla forces adhere to the principles of attrition, but simultaneously probe for weaknesses and pave the way for a concentrated attack. Without a regular force to augment the guerrilla forces, their task must be focused on inflicting cumulative damage piecemeal over a long period of time.

A conflict in which the phase of guerrilla warfare is successfully used in tactics is one in which "the local psychological atmosphere has become sufficiently unstable and the struggle tends to drag on forever. This is the classic form of endemic conflict and as a rule it can only be terminated by a compromise based on weariness" (Beaufre 1967, p. 117). As long as fighting is open and conducted by regular troops, there is a palpable enemy and the conflict is unlikely to become truly frozen. Guerrilla warfare in turn prolongs the fighting indefinitely, with the aim of exhausting the more powerful enemy. As Ho Chi Minh claimed, "we can surely fight for several years, till victory" (Cited in Giap 1968, p. 425).

Guerrilla warfare is thus essentially "a campaign of attrition" and the guerrilla "must hammer away constantly" (Guevara 1962, p. 11). This permanent and continuous attack is a feature that doesn't occupy such a prominent place in Mao's writings as in many others'. Mao wanted to develop the guerrillas into a regular army. Guevara seemed to take a more psychological approach to attacking the enemy and saw guerrilla forces as sufficient to win wars (Simpkin 1987, p. 25–27). If the "hammering away" is constant and surprises brought by the guerrillas a regular factor, it puts pressure on the enemy. He is robbed of time. There is no rest, not a moment when he can drop his guard and staying alert unceasingly erodes his ability to fight. Being battle-ready is exhaustive. "Is the enemy strong? One avoids him. Is he weak? One attacks him" (Giap 1962, p. 48). Attrition is not restricted to physical diminution of the enemy forces in skirmishes, but continues on the mental level. There is a constant threat of a surprise attack erupting at any given moment, anywhere. In such war there should be "no fixed line of demarcation, the front being wherever the enemy is found" (Ibid.). The unrelenting pressure and stress drain the morale and fighting spirit of the enemy. The Vietnamese way of warfare attempted to

> coordinate our activities in various operational areas, continuously attack the enemy everywhere and at any given time, sow chaos in his rear base, and

> direct powerful blows at him. We can defeat the enemy in a protracted war. We also possess conditions to create opportunities, bide for time, direct timely, vigorous blows at the enemy, and win increasingly great victories. (Giap 1970, p. 63)

Time was on the side of non-conventional forces and attrition worked on their behalf. Setting the rhythm of operations occurred in contradictory terms to those of the U.S. forces. The Viet Cong aim was to allow mental attrition of the enemy to secure their victory. "Our strategy was, as we have stressed, to wage a long-lasting battle" (Giap 1962, p. 27). The idea of seizing opportune moments was shared by both belligerents, but Giap's troops bided for time and waited for the best moment to act. No matter how asymmetric a war is, it still has to end in victory. This means that while the war may seem defensive, it is only a temporary phase until conditions have been created for the offensive. For Giap the entire "revolution is an offensive" and offensive should be the ideological basis of strategy (Giap 1970, p. 64).

Thus, the guerrilla strategy relies on accumulation of small effects and avoidance of decisive battle (Wylie 2014, p. 54). Even more importantly, there is no place in guerrilla operations for defense. While an attack should be always launched and pushed with the greatest haste, the actual fighting may last several days, if the enemy is to be annihilated (Mao 2000, p. 97). If the guerrillas took a defensive stand, they would lose the initiative. "When an army loses initiative, it loses its liberty; its role becomes passive; it faces the danger of defeat and destruction" (Mao 2000, p. 98). The guerrillas need to "consider the situation and decide at what time and at what place they wish to fight" (Mao 2000, p. 103). By choosing carefully when and where to retaliate, the guerrilla re-seizes initiative time after time. But making the choice depends on his ability to avoid open battles. When the guerrilla deems the time and place inappropriate, he must "move with the fluidity of water and the ease of the blowing wind" (Ibid.).

Having discussed guerrilla warfare at length, let us quickly address another aspect of asymmetry. Luttwak illustrated how the paradoxical logic of strategy creates situations in which irrational actions can be undertaken to benefit from the confusion caused by the asymmetric nature of this choice. Logic of strategy doesn't always follow the linear logic of common sense (Luttwak 1987, p. 4–8). Gray (2012, p. 37) agrees with the idea that strategy is paradoxical and adds that it frequently is ironic. Creveld (1991, p. 120) has further noted that the paradoxical logic works both in space and in time. Just because it is to be expected, the shortest route may end up being the longest due to enemy activity. In time, an operation once tried with failure may succeed when tried again, just because it is unexpected. Similarly something that has been successful will result in failure when tried again. Perhaps this is one of the reasons why Schelling (1963, p. 3) called the science of international strategy "retarded." Strategic behavior often contradicts common sense and accepted rules (Schelling 1963, p. 18). But the purpose is to create a surprise and without attempts to surprise the enemy, according to Luttwak, the art of war is reduced to administration. To surprise is to prevent the enemy from reacting in time and with full force. To achieve this

> paradoxical choices may be justified. Violating commonsense notions of what is best, as the shorter route is preferable to the longer, as daylight is preferable to the confusions of night, as full and ample preparation is preferable to hurried improvisation, the worst option may deliberately be chosen in the hope that the unfolding action will for that very reason be unexpected by the enemy, find him unready, and therefore diminish his ability to react. (Luttwak 1987, p. 8)

It may be logical to do illogical things to create conditions favorable to surprise. But this can lead to situations where both sides attempt to outguess the enemy and create endless chains of 'he thinks that I think that he thinks that I think that and therefore I do this.' Bad options become the best ones if the enemy is unable to anticipate them. This leads to a definition of surprise as "not merely one factor of advantage in warfare among many others, but rather the suspension, if only brief, if only partial, of the entire predicament of strategy, even as the struggle continues" (Luttwak 1987, p. 8). On occasion doing illogical things is an efficient way of jostling the symmetry and equilibrium of the enemy. Despite attempts at rationality we are constantly influenced by our emotions, fears, prejudices, and numerous other incalculable personal factors. In asymmetric warfare these may even help manage time. Temporary suspension of logic wins time if it confuses the enemy.

Asymmetric operational art can be created with the help of this paradoxical logic through choosing difficult methods and modes of action over simple ones. Hannibal leading his troops and elephants over the Alps was not the easiest path, but it surprised the enemy because it was unconceivable that anyone would voluntarily choose such route. Choosing the line of least expenditure of energy creates the simplest operational plans that are simultaneously the easiest to execute. However, just because of their predictability, they are unlikely to work against a superior enemy. Choosing less expected lines of operation means waste of energy and the structure of operations becomes complex, execution more demanding, and objectives harder to accomplish, but it may result in a surprise. Aiming for unpredictability is recommended but the unexpected may become predictable in time just as the safe and easy choice in operational art may be the hardest. The most outrageous operational plans become orthodox modes of action with repetition.

> Once time is duly introduced as a dynamic element, we can recognize the logic in its totality as the coming together, even the reversal, of opposites. And this is a process manifest not merely in the fate of counterconventional choices intended to achieve surprise, which eventually become quite predictable, but rather in all that is strategical, in all that is characterized by the struggle of adversary wills. In other words, if the passage of time is relevant and the paradoxical logic of strategy assumes a dynamic form, it becomes the coming together, even the reversal of opposites. In the realm of strategy, therefore, a course of action cannot persist indefinitely. It will tend to evolve into its opposite. (Luttwak 1987, p. 18)

For the weak to maintain the position of David against Goliath is easy, but the stronger belligerent may have to adapt and adopt. As warfare is reciprocal activity and since both sides learn from the other, the longer war continues, the more similar the enemies become (Creveld 2008, p. 307).

> The fighting itself will cause the two sides to become more like each other, even to the point where opposites converge, merge, and change places. Weakness turns into strength, strength into weakness. The principal reason behind this phenomenon is that war represents perhaps the most imitative activity known to man. The whole secret of victory consists of trying to understand the enemy in order to outwit him. (van Creveld 1991, p. 174)

Instead of thinking of asymmetry through limited paradigms like the weak against the strong, asymmetry can be addressed as an independent intellectual approach to operational art. In the planning phase of operations one should seek opportunities to exploit asymmetry. Asymmetry at its best means countering every aspect of the enemy's operational art with something disruptive. There is a long Chinese tradition of using two different approaches to warfare; the orthodox (*zheng*) and unorthodox (*qi*) (Yuen 2014, p. 34–35). Orthodox refers to conventional stratagems such as frontal attacks and unorthodox to unconventional and indirect approach (Scobell 2011, p. 201–205). A commander uses asymmetry advantageously when he has many different options to choose from.

> The greatest expression of destructive efficiency in tactical operations is combined-arms warfare. The commander who comes at the enemy with a variety of threats and attack profiles can probe him, search for weaknesses, and attack that weakness before the enemy can adapt. (Leonhard 1998, p. 136)

In other words joint operational warfare can be asymmetric if the commander chooses out of his assets the ones that exhibit and exploit asymmetry in the given situation. An airstrike is an example of highly asymmetric weapon if one possesses dominance of the air. Beyond choosing the right variety of capabilities to be employed, he should pay attention to choice of his tempo. If the enemy is rapid, he has to be countered with slowness to decelerate his operations. If the enemy employs advanced technology, agrarian methods will pose difficulties for him. If he wishes to win quickly, prolonging the war becomes profitable. One has to apply asymmetry also to the conception of time and temporality in warfare. On occasion, the paradoxical logic of strategy enacted through asymmetric and variable operational art means that time can be saved even by wasting it. It depends on the moment and the situation. Acting with variability can combine the unchanging and the momentary into a concoction most suited for the moment. Nothing should be permanent but

> tactics and procedures must remain in a state of flux, while the principles of maneuver warfare remain immutable. Attendant to the idea of adaptable tactics is the concept, repeatedly emphasized throughout Sun Tzu's works, of attacking weakness rather than strength. (Leonhard 1991, p. 34)

A society waging war benefits from its ability to change the paradigms of its doctrines to contradict those of its enemy. Cyber-attack against an enemy without computers would be useless but generally a new and technologically sophisticated operational art has shock-value. If both belligerents employ similar methods and technologies, the value decreases. In the West armed forces seem bent to follow the newest trends in doctrines and conventions of military hype. What if another path were chosen? Not a return to tribal and barbaric warfare per se, but what if operational art was consciously created to favor methods of the past in an attempt to find something so old that it becomes novel in contemporary context? Perhaps more violent, certainly less sophisticated, but able to cause discord in the enemy because asymmetry makes the two paradigms irreconcilable.

The history of war shows that civilizations formed around different developmental paradigms have huge difficulties in waging war on each other. It is by no means automatic that the more advanced society had an edge on the less developed and sophisticated. When Attila the Hun ravaged Europe or Genghis Khan's horde of horsemen over-rode everyone on their way westwards, they were on an inferior level of development. The Mongols were nomadic hunter-gatherers when they devastated agrarian cultures and the Huns like the Goths who laid much of the Roman Empire to waste were barbaric. Genghis Khan's campaigns serve as lessons in maneuver theory. With his horse armies "Khan was able to attain a degree of tactical and operational mobility unparalleled even by today's mechanized armies" (Leonhard 1991, p. 35). His methods were summarized by a survivor: "They came, they sapped, they burst, they slew, they plundered and they departed" (Juvaini 1958, p. 107). While efficient, this is not civilized warfare, but the army of Genghis Khan reflected the nature of the Mongol society (Smith 2008, p. 24). And indeed, since the Mongols left no political, cultural, or intellectual footprint, as Keegan (1998, p. 38) wrote, "they ended as they began, an army on horseback."

Similarly Arab warfare under T.E. Lawrence was successful because he didn't imbue them with the spirit of the indust-reality but freed the tribes to fight in their traditional way. By attacking the industrial base of war, such as railways, their battles stretched the Turks on the operational level. Smith called their methods an "antithesis of interstate industrial war" (Smith 2008, p. 167–168). The most advanced network society of today, the U.S., was in enormous trouble fighting the tribal and agrarian Taliban in Afghanistan just as it suffered in the heyday of air-mechanization against the peasant army of Viet Cong. In all spheres of warfare the idea of 'like fights like' is dangerous. It is a part of asymmetric nature of warfare that less developed societies can confront their opponent most successfully if they adhere to the ways and means that suit them best. These include cultural understandings and perceptions that combine to create a culture-specific way of interpreting and applying the principles of war in operational art. Thus, a technologically more advanced nation should use its assets fully in warfare and the less advanced nation benefits from using less sophisticated methods that are, nevertheless, characteristic to it.

Asymmetric operational art can be a huge cultural asset. Between an agrarian society and a techno-centric network society the paradigms of the art of war are so dissimilar that warfare becomes asymmetric. Networked information societies are hell-bent on saving time, being quicker, accomplishing more, and thus still following the indust-real rhythm. What if a different approach to time were chosen? Could one force the enemy battle rhythm offbeat? If the rhythm were more relaxed and more unpredictable, would it not become harder for the enemy to anticipate actions when there would be a variability of pace inbuilt to operations? In conflicts like Afghanistan or Vietnam one belligerent has the watches and the other has the time. It is not only the irregular enemy that may adhere to different timing that its opponent. Russians and Americans have fundamentally different conceptions of time in their military thinking. The U.S. supposedly is anchored in the here and now but Russia is culturally orientated toward the past and the distant future (Adamsky 2010, p. 51). If temporality is perceived differently in strategy, this manifests itself in operational art as well. Getting inside the enemy's decision-making loop is useless, since the loops revolve at different speeds and create different outputs. Any attempt to distort the enemy's rhythm is futile until this is understood.

3.9 Surprise as Operational Time-Robbery

Asymmetry manifesting itself through operational variability is effective in creating surprises. "Doing the same thing many times till the enemy is accustomed to it and then suddenly doing something quite different at the same time of day is sometimes effective in securing surprise" (Wavell 1953, p. 132–133). Despite its simplicity, this advice remains true. Meir Finkel (2011, p. 1) noted that military research has always focused on creating strategic surprise attacks, but claimed that the importance of doctrinal and technical surprises is increasing. The commander should always seek new methods of operational art and "surprise should be regarded as the soul of every operation. It is the secret of victory and the key to success" (Fuller 1926, 272). The principles of war should not be blindly obeyed since the challenge for the operational artist lies in comprehending each moment and ability to initiate action at *kairos*-times. Full clarity of the situation at all times through the fog of war is beyond reach but timely action on even a glimpse turns warfare into art. Highly complex thinking about operational art should result in simplified plans and these lead to simplified operations.

> Every complex operation takes time; and this time must be available without a counter-attack on one of its parts interfering with the development of the whole. If the enemy decides on a simple attack, one that can be carried out quickly, he will gain the advantage and wreck the grand design. So, in the evaluation of a complex attack, every risk that may be run during its preparatory stages must be weighed. The scheme should only be adopted if there is no danger that the enemy can wreck it by more rapid action. Wherever this is possible we ourselves must choose the shorter path. (von Clausewitz 1989, p. 228–229)

When one is absorbed in lengthy preparations a complex operation requires, the enemy is given a head-start. As a guideline, one should plan simple battles and operations that are quick to execute and build the element of surprise by combining these simple actions into more complex patterns on the levels of operational art and strategy. Mechanized forces prefer fast and simple operations. Martel wrote about the early deployment of tanks that the element of surprise was

> not so much in the method of attack as in the fact that the enemy could never tell when or where the assault would take place. In every case the utmost simplicity was aimed at, and this included preliminary training between infantry and tanks, whenever possible, so that everyone knew exactly what he was required to do. These were the keynotes of our success—simplicity and surprise. (Martel 1931, p. 35)

Surprises in maneuver warfare require speed and tasks, orders, and procedures need to be short and succinct. The simpler these are the less time is wasted in execution. Too often different stratagems only increase the difficulty of the operation. Luttwak (1987, p. 11) claimed that with the use of "any form of paradoxical action, notably secrecy, deception, and maneuver, the action will tend to become more complicated and more extended, thereby increasing the organizational risks in proportion." One must evaluate carefully how much time and benefit could potentially be gained from a surprise and how much time is lost with complications and if the surprise fails. Commander must decide based on his *coup d'oeil* whether to be simple, fast, and likely able to execute the plan or complicated, slower, and unsure of success in putting the plan to action. It depends upon the context which method would win time and which would squander it needlessly. Ultimately only the clash of arms judges whether the choice was right or not.

As Clausewitz wrote, everything in war is simple. Armed forces themselves have become more complex during the Second and Third Waves and so have the plans and orders needed to set them in motion. There are so many simple issues involved in warfare that their combination becomes excessively complicated (von Bernhardi 1914, p. 201). This is an important issue to tackle, since "the more complex the system, the greater the probability of failure" (Simpkin 1985, p. 179). But if complexity is inescapable, simplicity is an outdated objective since the traditional logic demanding that warfare should be simple "collapses in modern warfighting, (…) Information Age warfare demands not simplicity, but rather simplification—a completely different idea" (Leonhard 1998, p. 170). Every effort to simplify the complex is needed to conquer the friction of war.

The level of development in Third Wave societies is too high for Clausewitz's ideal of simplicity in warfare to remain valid. War is still continuation of politics with additional means but warfare has a different outlook. The ideal form of war remains unchanged but today "real warfare is not simple. Quite the reverse: It is almost inconceivably complex. The art and science of war demand a continuous process of analysis and simplification. We do not want simple plans; we want complex, effective plans that are simplified for execution" (Leonhard 1998, p. 176). Simplification can be a means of accelerating speed to surprise the enemy in accordance with tradition.

> Again, every single thing that must be done to supply, maintain, operate, and command the armed forces may be very simple once all the intangible mysteries that make combat possible are excluded. Yet together they are so complicated that the natural state of military forces of any size is chaos and immobility, from which only discipline and leadership can rescue any purposeful action. (Luttwak 1987, p. 11)

If warfare was only a question of planning it would still be possible to take note of all these issues and include them in the plan. But in warfare vast armies composed of individuals operate guided by their personalities, skills, fears, limitations, and passions. Only energetic command can turn this mess into efficient and purposeful action—and then the enemy has final say. To be able to execute his plan at all, the commander must overcome the friction and immobility by the force of his will.

Suvorov's military thought promoted strong leaders, and striving for surprise as main military virtues and surprise required speed. Thus, "to impose one's will on the enemy it is necessary to surprise him. The school of Suvorov teaches the art of achieving surprise and of securing one's own troops against it. Surprise is obtained above all by speed—speed in conceptions and in execution" (Andolenko 1956, p. 15–16). Simultaneously striving for surprise and seeking to secure oneself from being surprised in turn require forethought. In contemporary warfare the possibilities and chances of the unexpected occurring are so numerous that all of them cannot be eliminated. Imagination and forethought couple with flexible adaptation to events. When commanders stick to the 'tested and true' they leave themselves open for surprises. Assuming that the enemy is predictable creates a situation where having to counter novel tactics creates operational level surprises. This led Sikorski (1943, p. 139) to write that "in modern war, surprise would be the rule in planning a battle." We find this same emphasis in an ancient Chinese dialogue.

> King Wei said: 'Is there a Way (Tao) for one to attack ten?'
> Sun Pin said: 'There is.' 'Attack where they are unprepared; go forth where they will not expect it' (Sun Pin 1995, p. 89)

Tactical surprises are the rule in warfare but surprises are, "infinitely more important and effective in strategy than in tactics. Tactical initiative can rarely be expanded into a major victory, but a strategic one has often brought the whole war to an end at a stroke" (von Clausewitz 1989, p. 363). Tactical initiative can only win a battle but this feat has to be achieved several times in succession to emerge victorious from the operation and the war. Zhukov wrote about "operational surprises" referring to the ability of Soviet troops to concentrate their forces and prepare an attack sooner than expected. This type of surprise was confined to theater-level operations (Zukov 1969, p. 471, 594–595). In operational level, surprises are often caused by time management, but the results of imaginative operational art, however, are likely to be of great consequence on the strategic level.

As Clausewitz (1989, p. 350–351) wrote surprise is "the most important element of victory. Napoleon, Frederick II, Gustavus Adolphus, Caesar, Hannibal, and Alexander owe the brightest rays of their fame to swiftness." Swiftness was the key to surprise because the enemy is robbed of his time to react and prepare. Planning must search for the chance of surprising the enemy and possibilities for it are endless when imagination guided by reason and supported by sufficient mobility is allowed to roam freely. As Fuller wrote (1926, p. 272), "in war surprise in omnipresent; wherever man is there lurks the possibility of surprise." Spontaneous action in seizing the initiative and grasping a *kairos*-moment creates tactical surprises, but forethought, planning, and preparation are often required for an operational surprise.

In operational art, Fuller (1926, p. 277) argued, the means of creating surprise consist of the trinity of simplicity, speed, and secrecy of movement. For Liddell Hart (1927, p. 218–219) similarly "mobility and manoeuvre [are] the means to surprise in time and space respectively." Again, 'mindless movement' is insufficient and a lot of forethought is required to create simplified and efficient maneuvers. Simplification demands that the planned maneuver cannot be too intricate to avoid confusion and congestion and it enables fluent execution to save time. Secrecy of maneuver deprives the enemy of reaction time, and speed ensures unpreparedness of the enemy.

Finkel (2011, p. 47) argued that strategic importance of a surprise attack has decreased. As armies progressed from the First to Second and then the Third Wave, the chances of strategic surprises indeed seem to have diminished dramatically, but they have not vanished completely (Ehrfurth 1991, p. 381). In 1914 the German army surprised the French with the Schlieffen Plan. Germans planned to re-enact essentially the same plan in the advent of WW II, until Hitler intervened. As Field-Marshal Keitel described it,

> Hitler turned to us and said something like: 'That is just the old Schlieffen plan, with a strong right flank along the Atlantic coast; you won't get away with an operation like that twice running. I have quite a different idea and I'll tell you (i.e. Jodl and myself) about it in a day or two and then I'll talk it over with the War Office myself.' [...] it was Hitler himself who saw the armoured break-through at Sedan, striking up to the Atlantic coast at Abbeville, as the solution; we would then swing round [northwards] into the rear of the mortised Anglo-French army, which would most probably be advancing across the Franco-Belgian frontier into Belgium, and cut them off. (Keitel 1966, p. 102–103)

We should not rush to credit Hitler for the final execution. While a benefit of reusing the Schlieffen Plan would have been simplicity of execution, there was nothing novel or surprising about it and thus thoughts turned to another direction. As Manstein phrased it, the original idea for the offensive

> struck me as being essentially an imitation of the famous Schlieffen Plan of 1914. I found it humiliating, to say the least, that our generation could do nothing better than repeat an old recipe, even when this was the product of a man like Schlieffen. What could possibly be achieved by turning up a war plan our opponents had already rehearsed with us once before and against whose repetition they were bound to have taken full precautions? (von Manstein 1982, p. 98)

Hitler discredited the old version, but Manstein was the father of the plan to use strong armored troops to attack the Maginot Line in the vicinity of Sedan via routes through Luxembourg and southern Belgium. After penetrating the fortified line the attack was to be expanded (Guderian 1956, p. 81–82; von Manstein 1982, p. 94). In the end the only similarities between the plans were placement of the main force on the northern wing, the march through Belgium, and the mistake of letting operational art gain priority over strategy (von Manstein 1982, p. 99; Gray 2007, p. 107). Schlieffen opted for strategic surprise, and planners in WW II managed to create conditions for operative surprise. In both cases Germany had an inventive method of attack that almost led to the collapse of Western Europe (Heuser 2002, p. 62).

The meaning of surprise can be concentrated into Fuller's argument that "the power of surprise lies in stunning the reason. Men have no time to think: shall we do this, or shall we do that? Leadership on these occasions is frequently reduced to zero" (Fuller 1926, p. 321). A surprise is aimed at the commanders. Naturally every soldier caught in a whirlwind of the unexpected is influenced and his ability to function reduced, but instinct takes over and he adapts sooner or later. Time spent in finding a way to cope is less crucial in the case of a soldier than that of an operational commander. If he has no time to reason and his control slackens, every passing moment adds to the confusion. The goal of surprise is to

> attack the will of the enemy by accentuating fear, for, if a man is reduced to such a state of fear that he can do nothing save think of protection, he is at our mercy, for his moral endurance has ceased to dominate him. (Fuller 1926, p. 273)

Commanders try to plan for every eventuality and their plans attempt to expect the unexpected and unfavorable. There are innumerable ways to create surprises but only a limited amount of different types of surprises. Sikorski divided them into three categories of strategical, tactical, and technical surprises. We might add operational surprise to the list. As Sikorski (1943, p. 138) argued, "the ideal thing, or course, would be to achieve complete surprise in every sphere of action." In reality a surprise in even one category is a worthy goal since

> there is one important military principle of almost eternal validity: if, at the beginning of a war, absolute numerical superiority is not obtainable, one should try to be superior at least in one important weapon. (Ehrfurth 1991, p. 545)

Technical surprise has a profound impact on possibilities of other types of surprises but, as we have seen, it has short duration. Operational and tactical surprises are caused by new and original ways of warfare. In all types of surprises the common denominator is the ability to be unexpected. This may occur because of a new doctrine, which often results from technical innovation that enables novel actions. "Technical surprise has always had the greatest influence on the evolution of war in accordance with the importance and the efficiency of the new armaments and of the inventions relating to them" (Sikorski 1943, p. 140). New innovations result in new armaments and they require new tactics. Since proof of surprise is its result, it doesn't matter if it occurs because of a new weapon or a new warfighting doctrine (Finkel 2011, p. 27). The two go hand in hand.

Early stages of war are replete with auspicious moments for technical surprises but new technologies can have operational surprise effect at any phase of war. For example, the roll-out of tanks in Cambrai in 1917 had potential for a decisive operational victory with strategic repercussions, if only the tanks had been deployed more daringly and *en masse*. This was a finding of the British tank proponents, but shared by the Germans as well (Goodspeed 1966, p. 229–230). An operational breakthrough of the German lines had been "well within the bounds of possibility, if only the tanks had been employed in a more effective way, and if the offensive tactics of the other arms had been brought into full accord with the performance of the new arm, the tank" (Guderian 1992, p. 67). As there were insufficient mobile reserves, the attack wasn't continued into the depth and it lacked close-air support (Guderian 1992, p. 89). It speaks volumes about the pace of development that by April of 1918 tanks were so widespread among the belligerents that the first battle pitting tanks against each other occurred as the Germans attacked Villers Bretonneux (Martel 1931, p. 30). When the tank was no longer a novelty item on the battlefields the primary means to surprise were to be found in the rapidity of maneuver the tanks epitomized. Therefore,

> possessing both strategic and tactical mobility, mechanized forces may be speedily concentrated and employed. More so than in the case of other ground forces, this characteristic inherent in the mechanized arm permits of surprise employment. The preparations for the attack must aim, therefore, at surprise effect. (Guderian 1937, p. 11)

The ability of mechanized units to pulsatingly disperse and concentrate like an ebb and flow of force and redirect their effort can create favorable conditions for operational surprise. Planning and preparations to ensure fluidity of movement and maneuver constitute a precondition of success.

> Surprise may be attained through speedy and well-concealed movements, through the appropriate preparation and execution of the attack, and through new weapons of unprecedented capability. The rapid execution of the armoured attack is of decisive importance to the outcome of the battle. (Guderian 1992, p. 205)

By the time WW II started, the tank was a well-established weapon and could create no technical surprise. Thus Guderian focused on employing the tanks as weapons instead of infantry protection. On an existing technology he built new tactics. In this sense the result was what Finkel (2011, p. 2829) labeled "operational-tactical doctrinal surprise." New technology is useless unless coupled with a novel doctrine, and likewise the new doctrines require a mass of new materiel.

As Svechin argued the mass production of new weapons has to commence without testing them first on the battlefield. While this involves huge risks, it must be taken for surprise to be effective since "there is no need for gradualists, and experiment on the battlefield" (Svechin 1992, p. 128). Nevertheless, any new weapon is only auxiliary and used to support the existing weapons to supplement their effect. If one develops a doctrine based on massive use of a new type of weapon, this will sooner or later be countered by traditional but adapted operational art because it works on a wider spectrum of ways and means. There are, of course, different views. Leonhard (1998, p. 122–123) argued that in the future technology advances so fast that

> Even the idea of "wartime production" is outdated (…) Instead the future of material acquisition will be the rapid development and fielding of prototypes. The overwhelming numbers of the Sherman tank will be displaced by the dislocating quality of tomorrow's weapons.

Even if these weapons will be prototypes, to be effective, there must be as many of them as it is possible to produce in time. Mass production is cumbersome and Third Wave armies are so highly networked that every new weapon or technology implemented changes the composition of the entire system of systems (Kagan 2006, p. 41). With the addition of every new gadget the system needs to be reconfigured, and the more complex the system, the more widespread the reverberation of any new component. A nation unable to compete technologically can revert back to simpler and cheaper mass-produced materiel and weapons and fight asymmetrically.

Fuller (1926, p. 273) argued how all means of surprise ultimately "spring from the ability of the general, the courage of his men, and the perfection of their weapons." None of these are constants over longer periods of time. The courage of men may be bolstered or it may vanish, the superiority of any given weapon is passing, but the most obscure factor is the ability of the commander. "It is the dark horse of the battlefield. Mental ability is not so much a natural gift, save in the case of very few, as the product of scientific study—a close reasoning out of the values of conditions and an intelligent application of the principles of war" (Fuller 1926, p. 273).

To seize initiative and effect surprise to rob time in operational art one must combine the orthodox with the new and original. The traditions of the army need to be upheld and maintained for creating inner cohesion, but they should not be allowed to permeate operational art. "However praiseworthy it may be to uphold tradition in the field of soldierly ethics, it is to be resisted in the field of military command" (Rommel 1953, p. 204). In ethics and *esprit de corps* the tested and true remain a constant source of motivation. In operational art they are problematic. Only if the new and unorthodox cease to be effective, then the traditional and orthodox must be employed to stabilize the situation until some other scheme is developed.

> The danger to the professional soldier is that he inevitably tends to base his tactical thought on the methods which prevail at the moment, and with which he is in constant contact. He concentrates upon the attainment of mechanical perfection in the executive acts with which he is familiar. (Liddell Hart 1927, p. 191)

An attempt to perfect the mechanical execution of existing dogmas doesn't create operational art but only processes and procedures. "Prejudice against novel methods is a phenomenon typical of an officer corps raised in a proven system. (...) Everything other than the rule was a game of chance; and it followed that success could only be the result of luck or accident. This attitude leads to fixed ideas, the consequences of which are incalculable" (Rommel 2003, p. 96). As long as familiar methods are practiced and employed there will be no surprise, no sudden shock. Even if each maneuver is flawless, operational art becomes predictable and easily curbed by counter-measures. Military genius has another way of working. True genius "transcends mere copying; he refuses to swim with the stream; he strikes out in a direction of his own; and, what appears almost a miracle to the crowd, he frequently succeeds in diverting the stream from its course by compelling it to swirl forward in his own direction" (Fuller 1926, 98–99). The genius produces something novel and likely to astound the enemy.

Unfortunately, all too often commanders are confronted with another type of surprise. "On the battlefield itself a general is frequently surprised by his own stupidity, his lack of being able to appreciate conditions or apply to them the principles of war" (Fuller 1923, p. 38). It is a direct byproduct of the thoroughness in operational planning. The meticulous nature of the contemporary planning processes creates an illusion that every eventuality has been covered. When things develop in an unforeseen direction the plan collapses. The higher the level of detail in the original, the longer it takes to amend it. "The main causes of surprise are lack of foresight, loss in sensing the reality of war, lack in appreciating tactical values, and, above all, the standing grip of tradition which is ever choking our intelligence" (Fuller 1926, p. 281). Imagining that one is prepared for everything forebodes losing.

We have spent so long in discussion of surprise and its role in time management because it is the most effective way of forcing discord into the enemy rhythm to gain more time to act. To create a surprise one must be able to manage time and rob the enemy of his. "I call war the science of robbery, and not that of murder, as it has been hitherto denominated; because, to rob is its object, and to kill is only a means" (von Bülow 2013, p. 226). One can use the time 'robbed' to benefit his operations and grow the relative advantage over the enemy. Surprise is instantaneous, but its effects echo and accumulate over time. As Leonhard (1998, p. 184–185) wrote, "the condition of surprise cannot exist apart from the dimension of time. Surprise is defined in temporal terms, not physical. Fish live in water; surprise lives in time."

To sum up, this chapter focused on physical activities within the battlespace and especially the rhythmic pulsation from defensive to offensive and choosing the *kairos*-moments to alter the rhythm of the operation. This chapter argued that there are practical limits to the speed of operations reached long before the limits to potential velocity set by technology. In a world where armies focus on winning and managing time by being faster and faster, the cost-efficient way to win time is by being unorthodox and unpredictable in one's operational art through asymmetric methods and surprising the enemy. Seizing the correct moments to act and managing the tempo are crucial time-related issues concerning all activity within the battlespace. Asymmetry of timing, thinking, and tactics can be a valuable asset but the full employment of asymmetry requires the ability not only to use the newest technologies and methods, but to be prepared to revert even to old methods originating from a different age if they enable forcing the enemy of his rhythm. While the concepts of rhythm, tempo, and asymmetry are essentially intellectual and mental, they are manifested within the battlespace in concrete activity. The operational artist can win and control time either through energetic action, intellectual activity, or a combination of both. Thus, the next chapter looks at how time can be won alongside physical activity by using intellect, energy, and imagination as means to master time. It discusses ways of thought instead of modes of action and focuses on the mental processes, personal traits, and perceptions involved in operational art.

References

Adamsky, D. (2010). *The Culture of Military Innovation – The Impact of Cultural Factors on the Revolution in Military Affairs in Russia, the US, and Israel.* Stanford: Stanford Security Press.

Alberts, D. S., Garstka, J. J., & Stein, F. P. (2000). *Network Centric Warfare – Developing and Leveraging Information Superiority* (2nd ed. (revised)). Washington, D.C.: DoD C4ISR Cooperative Research Program.

Andolenko, S. (1956). The Imperial Heritage. In B. H. Liddell Hart (Ed.), *The Red Army* (pp. 13–23). New York: Harcourt, Brace and Company.

Bassford, C. (1994). *Clausewitz in English: The Reception of Clausewitz In Britain and America 1815–1945*. New York/Oxford: Oxford University Press.

Bassford, C., Moran, D., & Pedlow, G. W. (2010). *On Waterloo – Clausewitz, Wellington, and the Campaign of 1815*. Clausewitz.com.

Beaufre, A. (1965). *An Introduction to Strategy – With Particular Reference to Problems of Defense, Politics, Economics, and Diplomacy in the Nuclear Age*. New York/Washington: Frederick A. Praeger.

Beaufre, A. (1967). *Strategy of Action*. New York: Praeger.

Bond, B. (1998). *The Pursuit of Victory: From Napoleon to Saddam Hussein*. Oxford: Oxford University Press.

Caesar, G. J. (2010). *De Bello Gallico and Other Commentaries*. Miami: HardPress Publishing.

Cleary, T. (2005). *The Japanese Art of War – Understanding the Culture of Strategy*. London: Shambhala Classics.

Coker, C. (2010). *Barbarous Philosophers: Reflections on the Nature of War from Heraclitus to Heisenberg*. New York: Columbia University Press.

Colby, E. (1943). *Masters of Mobile Warfare*. Princeton: Princeton University Press.

Coram, R. (2002). *Boyd, The Fighter Pilot Who Changed the Art of War*. New York: Back Bay Books.

Corbett, J. (1999). Some Principles of Maritime Strategy. In *Book 4: Roots of Strategy* (pp. 149–262). Mechanicsburg: Stackpole Books.

Corum, J. S. (1992). *The Roots of Blitzkrieg – Hans von Seeckt and German Military Reform*. Lawrence: University Press of Kansas.

Danchev, A. (1998). *Alchemist of War – The Life of Basil Liddell Hart*. London: Phoenix Giant.

De Gaulle, C. (1976). *The Army of the Future*. Westport: Greenwood Press.

de Jomini, A. H. (1992). *The Art of War*. London: Greenhill Books.

de Jomini, A. H. (2007). *The Art of War*. Rockville: Arc Manor.

de Saxe, M. (1944). *Reveries Upon the Art of War*. Harrisburg.

de Saxe, M. (1987). My Reveries Upon the Art of War. In T. R. Phillips (Ed.), *Roots of Strategy* (pp. 177–300). Harrisburg: Stackpole Books.

Delbrück, H. (1990a). *Warfare in Antiquity – History of the Art of War* (Vol. I). Lincoln/London: University of Nebraska Press.

Delbrück, H. (1990b). *The Barbarian Invasions – History of the Art of War* (Vol. II). Lincoln/London: University of Nebraska Press.

Delbrück, H. (1990c). *Medieval Warfare –History of the Art of War* (Vol. III). Lincoln/London: University of Nebraska Press.

Delbrück, H. (1997). *Delbrück's Modern Military History*. Edited and Translated by Arden Bucholz. Lincoln/London: University of Nebraska Press.

Echevarria, A. J., II. (2000). *After Clausewitz – German Military Thinkers Before the Great War*. Lawrence: University Press of Kansas.

Echevarria, A. J., II. (2011). American Operational Art, 1917–2008. In J. A. Olsen & M. van Creveld (Eds.), *The Evolution of Operational Art – From Napoleon to the Present* (pp. 137–165). Oxford: Oxford University Press.

Ehrfurth, W. (1991). Surprise. In *Book 3: Roots of Strategy* (pp. 351–554). Mechanicsburg: Stackpole Books.

Finkel, M. (2011). *On Flexibility – Recovery from Technological and Doctrinal Surprise on the Battlefield*. Stanford: Stanford University Press.

Foch, F. (1920). *The Principles of War*. New York: Henry Holt and Company.

Foch, M. F. (1931). *The Memoirs of Marshal Foch*. Garden City: Doubleday, Doran and Company.

Foertsch, H. (1939). *Nykyinen ja tuleva sotataito* (Org. Kriegskunste Heute und Morgen). Porvoo: Werner Söderström Oy.

Franks, T. (2004). *American Soldier*. New York: Regan Books.

Fuller, J. F. C. (1920). *Tanks in the Great War 1914–1918*. London: John Murray.

Fuller, J. F. C. (1923). *The Reformation of War*. London: Hutchinson.

Fuller, J. F. C. (1926). *The Foundations of the Science of War*. London: Hutchinson.

Fuller, J. F. C. (1932). *The Dragon's Teeth – A Study of War and Peace*. London: Constable.

Fuller, J. F. C. (1946). *Armament and History – A Study of the Influence of Armament on History from the dawn of Classical Warfare to the Second World War*. London: Eyre & Spottiswoode.

Fuller, J. F. C. (1948). *The Second World War 1939–45. A Strategical and Tactical History*. London: Eyre & Spottiswoode.

Fuller, J. F. C. (1960). *The Generalship of Alexander the Great*. New Brunswick: Rutgers University Press.

Fuller, J. F. C. (1961). *The Conduct of War 1789–1961: A Study of the Impact of the French, Industrial, and Russian Revolutions on War and Its Conduct*. London: Eyre & Spottiswoode.

Giap, V. N. (1962). *People's War People's Army – The Viet Cong Insurrection Manual for Underdeveloped Countries*. New York: Frederick A. Praeger.

Giap, V. N. (1968). *"Big Victory, Great Task" – North Viet-Nam's Minister of Defense Assesses the Course of the War*. New York: Frederick A. Praeger, Publishers.

Giap, V. N. (1970). *Banner of People's War, the Party's Military Line*. London: Pall Mall Press.

Goodspeed, D. J. (1966). *Ludendorff – Genius of World War I*. Boston: Houghton Mifflin Company.

Gray, C. S. (2007). *Fighting Talk – Forty Maxims on War, Peace, and Strategy.* Westport: Praeger Security International.

Gray, C. S. (2012). *Airpower for Strategic Effect.* Alabama: Air University Press, Maxwell Air Force Base.

Guderian, H. (1937). *Armed Forces and their Cooperation with Other Arms.* Fort Humphreys: The Army War College.

Guderian, H. (1956). *Sotilaan Muistelmia*, Org. Erinnerungen Eines Soldaten (trans: Halsti, W. H.). Helsinki: Kustannusosakeyhtiö Otava.

Guderian, H. (1992). *Achtung – Panzer! The Development of Tank Warfare* (trans: Duffy, C.). London: Cassell.

Guevara, C. (1962). *On Guerrilla Warfare* (2nd ed.). New York: Praeger.

Handel, M. I. (2001). *Masters of War – Classical Strategic Thought* (3rd revised and expanded ed.). London/New York: Routledge.

Heuser, B. (2002). *Reading Clausewitz.* New York: Random House.

Heuser, B. (2010). *The Evolution on Strategy – Thinking Strategy from Antiquity to the Present.* Cambridge: Cambridge University Press.

Howard, M. (1983). *Clausewitz.* Oxford: Oxford University Press.

Isserson, G. S. (2013). *The Evolution of Operational art* (trans: Menning, B. W.). Fort Leavenworth: Combat Studies Institute Press.

Juvaini, 'Ala-ad-Din 'Ata-Malik. (1958). *The History of the World-Conqueror* (Vol. I). Cambridge: Harvard University Press.

Kagan, F. W. (2006). *Finding the Target – The Transformation of American Military Policy.* New York: Encounter Books.

Kaldor, M. (2007). *New & Old Wars* (2nd ed.). Stanford: Stanford University Press.

Kautilya. *Arthashastra.* http://www.sdstate.edu/projectsouthasia/upload. Downloaded 7 Oct 2011.

Keegan, J. (1998). *War and Our World. The Reith Lectures 1998.* London: Hutchinson.

Keitel, W. F.-M. (1966). *The Memoirs* (trans: Irving, D.). New York: Stein and Day Publishers.

Lawrence, T. E. (1997). *Revolt in the Desert.* Hertfordshire: Wordsworth Edition.

Leo, VI. (2010). *Taktika.* Text, Translation and Commentary by G. T. Dennis. Dumbarton Oaks Texts XII, Washington, DC.

Leonhard, R. R. (1991). *The Art Of Maneuver – Maneuver-Warfare Theory and airland battle.* Novato: Presido.

Leonhard, R. R. (1998). *The Principles of War for the Information Age.* New York: Ballantine Books.

Liddell Hart, B. H. (1927). *The Remaking of Modern Armies.* London: John Murray.

Liddell Hart, B. H. (1932). *When Britain Goes to War: Adaptability and Mobility.* London: Faber and Faber.

Liddell Hart, B. H. (1937). *Europe in Arms.* London: Faber and Faber.

Liddell Hart, B. H. (1946). *The Revolution in Warfare*. London: Faber and Faber.
Liddell Hart, B. H. (1950). *Defence of the West – Some Riddles of War and Peace*. London: Cassell and Company.
Ludendorff, E. (1919). *Sotamuistelmani 1914–1918*. Porvoo: WSOY.
Luttwak, E. N. (1987). *Strategy: The Logic of War and Peace*. Cambridge, MA: Belknap Press of Harvard University Press.
MacArthur, D. (1964). *Reminiscences*. London: Heineman.
Mao, T.-T. (1963). *Selected Military Writings*. Peking: Foreign Languages Press.
Mao, T.-T. (2000). *On Guerrilla Warfare*. Urbana/Chicago: University of Chicago Press.
Martel, G. L. Q. (1931). *In the Wake of the Tank: The First Fifteen Years of Mechanization in the British Army*. London: Sifton Praed.
Minh, H. C. (2008). *Prison Diary*. Hanoi: The Gioi Publishers.
Montgomery. (2000). *A Concise History of Warfare*. Hertfordshire: Wordsworth Editions Limited.
Moran, D. (2007). The Instrument: Clausewitz on Aims and Objectives in War. In H. Strachan & A. Herberg-Rothe (Eds.), *Clausewitz in the Twenty-First Century* (pp. 91–106). Oxford: Oxford University Press.
Münkler, H. (2005). *The New Wars*. Cambridge: Polity Press.
Musashi, M. (1995). *Maa, Vesi, Tuli, Tuuli ja Tyhjyys*. (Orig. *Go rin no sho*), Helsinki: Otava.
Napoleon B. (1987). *The Military Maxims of Napoleon* (Ed. G. David). London: Chandler, Greenhill Books.
Paret, P. (2015). *Clausewitz in his Time. Essays in the Cultural and Intellectual History of Thinking About War*. New York: Berghahn Books.
Patton, G. S., Jr. (1947). *War as I Knew it*. Cambridge: Riverside Press.
Ritter, G. (1958). *The Schlieffen Plan: Critique of a Myth*. London: Oswald Wolff Publishers Limited.
Rokossovsky, K. (1970). *A Soldier's Duty*. Moscow: Progress Publishers.
Rommel, E. (1953. *The Rommel Papers* (Ed. B. H. Liddell Hart). London: Collins.
Rommel, E. (2003). *Rommel and his Art of War* (Ed. J. Pimlott). London: Greenhill Books.
Schelling, T. C. (1963). *The Strategy of Conflict*. Cambridge, MA: Harvard University Press.
Schwarzkopf, N. H. (1993). *It Doesn't Take a Hero* – The Autobiography, written with Peter Petre. New York: Bantam Books Paperback edition.
Scobell, A. (2011). The Chinese Way of War. In J. A. Olsen & M. van Creveld (Eds.), *The Evolution of Operational Art – From Napoleon to the Present* (pp. 195–221). Oxford: Oxford University Press.
Sherman, W. T. (1890). *Personal Memoirs of Gen'l W.T. Sherman* (3rd ed., Revised and Corrected, Two Volumes, Vol. I., Ed. L. Charles) New York: Webster & Company.

Sikorski, G. W. (1943). *Modern Warfare – Its Character, Its Problems.* London: Hutchinson.

Simpkin, R. E. (1985). *Race to the Swift – Thoughts on Twenty-First Century Warfare.* London: Brassey's Defence Publishers.

Simpkin, R. E. (1987). *Deep Battle – The Brainchild of Marshal Tukhachevskii.* London: Brassey's Defence Publishers.

Smith, R. (2008). *The Utility of Force: The Art of War in the Modern World.* New York: Vintage Books.

Ssu-ma. (1993). The Methods of the Ssu-ma. In R. D. Sawyer (Ed.), *The Seven Military Classics of Ancient China* (pp. 107–144). Boulder: Westview Press.

Stempel, J. (2012). *The Nature of War – Origins and Evolution of Violent Conflict.* Jefferson: McFarland & Company.

Stoker, D. (2010). *The Grand Design – Strategy and the U.S. Civil War.* Oxford/ New York: Oxford University Press.

Strachan, H. (2013). *The Direction of War – Contemporary Strategy in Historical Perspective.* Cambridge: Cambridge University Press.

Sumida, J. (2007). On Defence as the Stronger Form of War. In H. Strachan & A. Herberg-Rothe (Eds.), *Clausewitz in the Twenty-First Century* (pp. 163–181). Oxford: Oxford University Press.

Sun Pin. (1995). *Military Methods. Translated with introduction and commentary by Ralph Saywer.* Boulder: Westview Press.

Sun-tzu. (1993). Art of War. In R. D. Sawyer (Ed.), *The Seven Military Classics of Ancient China* (pp. 145–186). Boulder: Westview Press.

Svechin, A. A. (1992). *Strategy.* Boulder: Eastview Press.

T'ai Kung. (1993). Six Secret Teachings. In R. D. Sawyer (Ed.), *The Seven Military Classics of Ancient China* (pp. 40–106). Boulder: Westview Press.

Thucydides. (1971). *The Peloponnesian War* (trans: Warner, R.). Harmondsworth: Penguin Books.

Triandafillov, V. K. (1994). *The Nature of the Operations of Modern Armies* (trans: Burhans, W. A.). Portland: Frank Cass.

Trythall, A. J. (1977). *'Boney' Fuller – The Intellectual General 1878–1966.* London: Cassell.

van Creveld, M. (1991). *On Future War.* London: Brassey's.

van Creveld, M. (2008). *The Culture of War.* New York: Ballantine Books.

van Creveld, M. (2011). Napoleon and the Dawn of Operational Warfare. In J. A. Olsen & M. van Creveld (Eds.), *The Evolution of Operational Art – From Napoleon to the Present* (pp. 9–34). Oxford: Oxford University Press.

Vego, M. (2009). *Joint Operational Warfare: Theory and Practice* (Revised ed.). Newport: U.S. Naval War College.

von Bernhardi, F. (1914). *How Germany Makes War.* London: Hodder & Stoughton.

von Bülow, D. H. (2013). *The Spirit of the Modern System of War.* Cambridge: Cambridge University Press.
von Clausewitz, C. (1989). *On War.* Edited and translated by Michael Howard and Peter Paret. First paperback edition. Princeton: Princeton University Press.
von der Goltz, F.-M. B. (1906). *The Nation in Arms – A Treatise on Modern Military Systems and the Conduct of War.* London: Hodder and Stoughton.
von Freytag-Loringhoven, H. (1991). The Power of Personality in War. In *Book 3: Roots of Strategy* (pp. 173–349). Mechanicsburg: Stackpole Books.
Von Leeb, R. (1991). Defense. In *Book 3: Roots of Strategy* (pp. 1–172). Mechanicsburg: Stackpole Books.
von Manstein, E. (1982). *Lost Victories.* Novato: Presido Press.
von Moltke, H. (1993). *On the Art of War – Selected Writings* (Ed. J. Daniel). New York: Hughes, Ballantine Books.
von Schileffen, C. A. (1936). *Cannae.* Fort Leavenworth: The command and General Staff School Press.
von Seeckt, H. (1930). *Valtiomies ja Sotapäällikkö.* Helsinki: Otava.
Warden, J. A. III (1995) The Enemy as a System. *Airpower Journal, Spring 9,* 40.
Warden, J. A., III. (2000). *The Air Campaign – Planning for Combat.* San Jose: to Excel.
Wavell, A. (1953). *Soldiers and Soldiering or Epithets of War.* London: Jonathan Cape.
Wu-tzu. (1993). *The Seven Military Classics of Ancient China* (Ed. R. D. Sawyer, pp. 187–224). Boulder: Westview Press.
Wylie, J. C. (2014). *Military Strategy – A General Theory of Power Control.* Annapolis: Naval Institute Press.
Yuen, D. M. C. (2014). *Deciphering Sun Tzu – How to Read the Art of War.* Oxford: Oxford University Press.
Zukov, G. Z. (1969). *Marsalkka Zukovin muistelmat.* Porvoo: WSOY.

CHAPTER 4

Winning Time Intellectually

To create conditions favorable for victories the commanders must combine their intelligence and imagination, originality and flexibility, audacity and boldness to master operational art and use time as both their ally and resource. There are numerous fleeting *kairos*-moments when initiating action is more auspicious than others and the operational artist and his plan must be flexible enough to identify and then exploit them. Time, tempo, and velocity may run amuck and the commander has to use his *coup d'oeil* to control them and use them to optimally support his operations. The inner eye is as important to producing art of war than the physical ones. Without intellect and imagination, there can be no operational art but only mechanical execution of principles and this was a source of constant irritation for Fuller (1923, p. vii) who wrote that "If I decry professional soldiers, it is not for personal dislike of an honest and modest body of men, but because their work is fatuous; they are not preparing for the next war but for the last war but one, and are consequently a danger rather than a security.

4.1 The Commander's Staff as a Timesaver

The commander-in-chief has a different function than operational artists. His task today on the strategic level is to confer with his political masters. He takes a detached stance from the battles and provides the objectives and the resources for the operational artists under his command (Smith

J. Hanska, *War of Time*,
https://doi.org/10.1007/978-3-030-45517-0_4

2008, p. x). The operational commander in turn must analyze the big picture and focus on combining battles into operations to fulfill his part in attaining strategic objectives by completing the operational tasks given to him. For the tactical commanders time management consists mainly of rapidity of decision and action. de Saxe wrote how the battle commander,

> when he sees an occasion, he should unleash his energies, hasten to the critical point at top speed, seize the first troops available, advance them rapidly, and lead them in person. These are the strokes that decide battles and gain victories. The important thing is to see the opportunity and to know how to use it. (de Saxe 1987, p. 296)

War in the agrarian age of de Saxe was far removed from its indust-real or Third Wave outlook and the general seldom commands his troops in person at the front. Nevertheless, the way to tactical and operational success for today can be found in this quotation. The general's mental capacity has to focus entirely on the conduct of operations. He needs to search for the advantageous time and place to turn the course of warfare and act right when the *kairos*-moment arrives.

One of the best advises of how to best utilize every hour at the disposal of the commander comes from the Tactics of Leo VI (2010, p. 571). His argument was that, "in time of war, during the night, plan what you have to do and, during the day, carry out what you have decided. For the same time is not suitable both for planning and taking action." No commander should be forced to plan ahead at while he is in charge of executing the current plan. This would be a schizophrenic task. The commander should focus on the single most important task at any given moment. Napoleon was the ultimate commander-in-chief and statesman who was able to dictate everything from battles to grand strategy. But more and more subordinates were involved in the process. The society and the military alike emerged from the revolution not only changed but also more complicated. This required alterations since the responsibility of operational commanders is

> an immense one where modern numbers are concerned. It is, indeed, seldom possible for a single man to fulfill it; several men are needed. This is the new conception which the French Revolution brought into war, by making the personal initiative of subordinate chiefs (all working in the same direction

> and complying with the same doctrine) concur in setting up a complete direction of armies. (Foch 1920, p. 290)

Staffs were invented to control this complexity. Even such an undisputed genius as Napoleon was unable to administer to every detail. Some responsibilities had to be delegated. Napoleon kept a tight leash of his staff and reserved the responsibility for all major decisions on himself. With the structure and organizational hierarchy of today's armies and especially the political structure above the commander-in-chief, things are no longer so clear-cut. The commander's hand cannot reach everywhere. No great captain of the past could sovereignly tackle all issues of today's operational art. The staffs need to be big and efficient enough to ensure not one single moment of the commander's time is wasted. Simultaneously they must be small enough to prevent their internal friction from causing extra delay to planning and decision-making.

One advantage of the staff is distribution of time to those functions that need it the most and where it best benefits the overall objective of the operation. Often this can be found at the top of the hierarchy. From the perspective of operational art any commander "who insists upon the writing and revision of his own orders, robs his mind of the leisure necessary for the conception of fresh ideas. He ought to think rather than wield the pen" (Goltz 1906, p. 46). If the commander gets tied down to repetitive and menial tasks, he lacks the luxury of freedom for creative thought. Theorists like Matheny (2012, p.19) or Michael Howard argued that the greatest military innovation of the nineteenth century was the general staff. The staff freed the commander to focus on operational art. All preparatory work belongs to the staff. They provide the commander with all the information he requires to allocate time and other resources where it is operationally most crucial.

The idea to use different parts of the staff to handle planning and execution was a product of the twentieth century. Ludendorff (1919, p. 45) still complained about the lack of rest and difficulties he had to face in fighting one battle to the end and simultaneously preparing the commencement of the next one. Modern massive staffs have enabled a part of the staff to plan the next phase of the operation while another part supports the commander in the conduct of combat at hand. Even then not every general is up to his task and its demands.

> Unless a man is born with talent for war, he will never be other than a mediocre general. It is the same with all talents; in painting, or in music, or in poetry, talent must be inherent for excellence. All sublime arts are alike in this respect. That is why we see so few outstanding men in any science. Centuries pass without producing one. (de Saxe 1944, p. 119–120)

An efficient staff is a necessity for the commander to carry out his tasks. Even if the commander is not an original genius, a professional staff can help him to create operational art. As geniuses are few and far between,

> the difficulty of always selecting a good general has led to the formation of a good general staff, which being near the general may advise him, and thus exercise a beneficial influence over the operations. A well-instructed general staff is one of the most useful of organizations. (de Jomini 2007, p. 41)

Amidst the ever-changing circumstances the commander has to understand their impact on the attainment of his objectives. Each minor surprise and change has to be evaluated according to how they affect the operational plan. If there is no significant impact, the execution of the plan can continue, else alterations are required. The staff closely follows the unfolding of events to be prepared to amend the operational plan accordingly. The commander has to remain above the minuscule details of the battle and freed to make informed operational decisions. This challenge of command has remained unchanged through the three waves. The devil is in the details that should not clutter the mind of the commander. As Vego (2009, p. III–19) phrased it, "time lost or wasted in conducting daily routine activities can sometimes indirectly influence, to a great degree, the outcome of a combat action. The commander has to focus and not take on too many diverse tasks so that he will not overlook matters essential to the battle" (Leo VI 2010, p. 291). "On the day of the battle, I should want the general to do nothing. His observations will be better for it, his judgment will be more sane, and he will be in better state to profit from the situations in which the enemy finds himself during the engagement" (de Saxe 1944, p. 118). This freedom from being preoccupied and overwhelmed with details optimally releases his intellectual capacity to observe the progress of operations. Fuller attributed much of Caesar's success to an ability to detach himself intellectually and emotionally from everything else than the combat. This

> clear-sightedness, which amounted to intuition, was largely due to his mind being undisturbed either by recollections or by expectations, and to the fact that he was seldom led away by illusions concerning the abilities of men or the appearance of events. He was a supreme opportunist, whose self-confidence, combined with his faith in his fortune, his audacity and his subtlety, enabled his to take chances unimagined by others. (Fuller 1965, p. 50)

The commander must be freed to lead and direct the conduct of operations. Staff takes care of administrative issues, while the operational commander focuses his energy "to subduing inevitable doubts, and to assuring absolute clearness of view. It is not wise to take upon himself too much of the details of execution, a fault to which small minds are very much inclined" (Goltz 1906, p. 46). This requires a mind of strong resolution. It is natural for thoughts to start to ramble when confronted with big decisions. The urge to descend to the level of details tempts the operational commanders because of their ease to comprehend and manage.

Undoubtedly commander-centric decision-making will prevail in the future. The shape of the staff structure is less certain. The Napoleonic staff was a small and nimble enabler of quick action. The current staffs of Western coalition forces are bloated and the friction their sheer size inserts into the decision-making cycle makes them inefficient. Hierarchy and bureaucracy were offspring of indust-reality that needed its hordes of coordinators and managers. Bureaucracy was an answer to how "the flow of decisions could be accelerated to keep up with the faster pace of life brought by industrialism" (Toffler 1990b, p. 143). In NCW theories there existed a tendency to favor less hierarchic and parallel command structures that signified a mental shift toward the Third Wave (Alberts et al. 2000). Some theorists, however, advocate growing the size of headquarters in the future in response to the supposed requirements of the information age. Since there is more information available, more people with more and more complex technological tools are required to manage information. "Technology will serve to harness information and arrange it to be of use to the commanders. But as technology of war becomes even more complex, military staffs need to adapt—and sometimes grow—in order to accommodate it" (Leonhard 1998, p. 199).

Bureaucratic command structures are ideal to solving routine problems and churning out administrative decisions at a preset pace in accordance with the principles of indust-reality. Acceleration of pace and corresponding compression of time coupled with non-routine problems push a bureaucratic system into at best a halt, or, at worst, utter chaos (Toffler 1990b, p. 139). With rapid changes more decisions need making. Every situation is more complex than the previous one and requires more information to create sufficient knowledge for a solution. A novel problem is more demanding to solve than one encountered numerous times before. The demand to gather and analyze more information in shorter time coupled with emergent problems creates a vicious decision-making cycle that accelerates out of control.

It becomes necessary at a certain point to slow the decision-making cycle. The speed accelerates automatically due to the urgent need for more data input and information output. Either some information-related processes, such as analysis, have to be automated or a new means of managing the decision-making cycle invented. Toffler (1990b, p. 125) proposed the adoption of a new organizational system for the Third Wave to supplant bureaucracy and called it "ad-hocracy". He meant this to be "the fast-moving, information-rich, kinetic organization of the future, filled with transient cells and extremely mobile individuals" (Toffler 1990b, p. 144). Such a command structure is ill-suited for the military. Besides, while it would bypass bureaucracy and streamline the chain of command, it would not ease the commanders' burden. No matter how well the staff prepares possible courses of action and proposes 'ready-made' options for the commander, the responsibility for the consequences rest on his shoulders. Decisions have to be made, and if the choices are too simplified, the room to employ the commander's *coup d'oeil* shrinks and the reason why the commander has been designated the command in the first place becomes meaningless. Maneuverability of troops today doesn't restrain the commander but the other way around. In operational art "top speed is often imposed by mental rather than muscular limitations. (…) the greater the number of alternative courses of action open to the subject, the longer it takes him to reach a decision and carry it out" (Toffler 1990b, p. 353). Attempts to increase speed of operations inadvertently decimate the quality of decisions.

Having arrived at the limits of what is humanly possible it is logical to discuss the aides of planning that information age offers to operational artists. The first and foremost one is the computer and in the near future artificial intelligence. Certain strictly specified parts of the decision-making process can be 'outsourced' to them. For Napoleon the staff officers of his day were what computers are to us today. They were reduced to machines that created the output required from them on command. Napoleon's centralized decision-making in practice limited the freedom of staff officers to use their powers of reasoning. Over time more responsibility and even demand for taking the initiative and making decisions was delegated to staff officers. Today the

> increasing reliance on computers as aids to planning is fraught with potential dangers. While the network of computers can considerably reduce the time needed to carry out routine planning tasks, it cannot replace the human mind and the skills and experience of operational commanders and their staffs. (Vego 2009, p. III–25)

While overt reliance on computers in planning tasks of today is a dangerous fallacy, the advances in their computing capacity, ability to solve more complex dilemmas, and artificial intelligence in the future will have an impact on the staffs. Nevertheless, no matter how well an artificial intelligence can calculate all eventualities of future warfare, the human commander will always have the chance to provide the game-changer of intuition and uphold the importance of operational art.

4.2 *Auftragstaktik* as Accelerated Decision-Making

In our journey from agrarian times up to contemporary outlook of warfare one thing becomes clear. The importance of time in achieving victory has grown, and increased intellectual effort is required from increased number of soldiers. The commander-in-chief or operational artists are no longer able to create their art unassisted. In the armies of ancient Greece, the citizen-soldier basically just did what he was told to do. Decision-making belonged to the commander-in-chief. In the agrarian age occasionally even his only task was to determine the battle formation and order the attack. Once the use of reserves became part of tactics and operational art, another decision-making point emerged. For Napoleon during every battle there is at least one moment when its tide can be turned and victory stolen. As warfare evolved the number of such *kairos*-moments has increased while the relative importance of each individual moment has waned in turn. Instead of one glorious clash

> battle has become a serial process composed of momentary minor opportunities, and the exploitation of these naturally tends to turn even more on a

> general superiority in minor tactics among the junior leaders than on the major tactics of the generals—except on the highest levels, where strategy is called for, and where a bad strategical decision can undo a tactical advantage.
>
> In other words, a battle has become a team-game on the largest scale, in which the junior leaders are players, not pawns. (…) The junior leaders have always borne the brunt of war, and do so still; but their intelligent initiative, and its cultivation, have now become vital factors in determining the issue. (Liddell Hart 1950, p. 315)

Battles are no longer clashes of phalanxes lasting at best a few hours. Now the duration is counted in days, perhaps weeks, and in this longer stretch of *kronos*-time more *kairos*-moments exist. As battles occur throughout the theater the boundaries of tactical and operational space and time are blurred. Therefore, lower-level commanders have to make decisions that influence the result of both the battle and the operation. Operational commanders are excluded from combat and their leadership focuses on operations. Junior leaders win or lose battles and provide victories for the operational commander to combine into an operation.

The German system of command and control, *Auftragstaktik,* that increased responsibilities of junior commanders didn't emerge from nowhere to suddenly revolutionize warfare. In one of his predictions of the character of future warfare de Bloch claimed that it was a normal evolution of the art of war that "of necessity the directing power must pass from the hands of the older commanders, not to speak of generals—from the hands of colonels and even commanders of battalions—into the hands of captains" (de Bloch 1914, p. 38). This was a logical phase in evolutionary process that had started in the antiquity. As closed formations became obsolete, soldiers had to increasingly use their initiative and make decisions themselves. Battles,

> now more than ever, are battles of men, of captains. They always have been in fact, since in the last analysis the execution belongs to the man in ranks. But the influence of the latter on the final result is greater than formerly. (du Picq 1987, p. 128)

Auftragstaktik is not only a tactical level method of command. Indeed, the lower down the military hierarchy we descend, the less room there is for applying it. A company commander must give highly detailed tasks to his platoon leaders. The commander-in-chief hopes to get only the order to win the war and full freedom to choose the means. The operational level commanders must be "fully acquainted with the strategic situation from which the action develops, the general and guiding plan of action must be briefly communicated to them. They must have learnt to understand from a few tersely coined words the idea and the will of the supreme commander. They must, however, not receive this will in the form of an order, but as a task, leaving them full liberty in the choice of the means for its execution" (von Bernhardi 1914, p. 107). In operational art *Auftragstaktik* frees tactical commanders to do what they deem best to do their part in attaining the operational objective. *Auftragstaktik* follows the principle that the one who carries the responsibility to fulfill the task must be allowed to choose freely the methods he wishes to use (von Seeckt 1930, p. 45). This allows for the creation of a flexible operational plan. Yet there is no doubt to whom responsibility for a victory or loss belongs. For Napoleon the commander

> is the head, he is the all of an army. The Gauls were not conquered by the Roman legions, but by Caesar. It was not before the Carthaginian soldiers that Rome was made to tremble, but before Hannibal. It was not the Macedonian phalanx which penetrated to India, but Alexander. (Cited in Fuller 1960, p. 281–282)

Foch echoed this idea and agreed that "history is therefore right in making generals responsible for victories—in which case they are glorified; and for defeats—in which case they are disgraced. Without a commander, no battle, no victory is possible" (Foch 1920, p. 288). The responsibility delegated to junior commanders is always temporary. It is offered when they are trustworthy and accepted when they trust themselves. In all their initiatives, they execute the will of their commander by attempting to fulfill his intent to the best of their abilities. The commander ultimately bears the burden or wears the laurel, which makes his intent crucial.

Planning is conducted regressively from the end-state set by the commander toward the present to determine the intermediate objectives to be reached within determined time limits to attain the final objective. Even when a plan has not been finalized in advance and one reacts to an utterly new situation under time-pressure, "though the time before the beginning of decisive movements preceding a battle is often very short, there will always be a few spare minutes in which to express the intention of the commander-in-chief in a sentence or two with clearness and precision"

(Goltz 1906, p. 61). Time can be spent wisely just as well as it can be saved stupidly. No matter how pressed for time, the commander's intent must be made clear so that the subordinate commanders can fathom his objectives and let these guide their own plans. The primary temporal focal point in operational art thus is the commander's intent of end-state he wants to accomplish and when. This is a prerequisite for any form of *Auftragstaktik*. "The intention of the commander-in-chief is the only guide regarding the conduct of subordinate commanders, if an order cannot be carried out in the way intended. It must therefore be made known to them with unmistakable clearness" (Ibid.). *Auftragstaktik* was an innovative idea of dispersing command that grew out of necessity to counter the discord and complexity of the battlespace where it may be impossible to contact subordinate commanders during the combat. "Troops have to rely, as long as the action lasts, solely on what they knew about the object of the action and the co-operation of the various units before the battle began" (von Bernhardi 1914, p. 104). Operational command at its best consists of

> clearly determining the result to be aimed at, the general function ascribed to each subordinate unit in the operation undertaken by the whole of the forces; at the same time such a determination must leave the subordinate chief entirely free to choose the means which have to be used in order to reach, in any particular case, the result demanded, and that in spite of adverse circumstances which cannot be foreseen in advance. (Foch 1920, p. 172)

The essence of *Auftragstaktik* is summarized here even prior to the experiences of WW I. The commanders couldn't be present in every decisive point in time to make all necessary decisions. They had to delegate authority to make decisions and there were two distinct ways to do this.

> A commander may indicate his decision to his subordinates either in the categorical form of a battle order, which indicates the situation in which it will be carried out, or in the form of a directive limited to a statement of the goals of operations for the next few days, which provides the executor with a great deal of freedom in choosing methods of achieving them. (Svechin 1992, p. 317)

Needless to say, the idea of *Auftragstaktik* is presented in the latter option that endows the subordinate commanders with freedom in contrast to binding orders. Terminology varies, but the essence of this method of command remains. The commander voices his intentions for the outcome and tells his subordinates what he wants to achieve operationally. Then he describes the role each unit plays in the whole and assigns them objectives. Speed of action and maneuver can be increased by creating a command structure and a method of leadership that supports them. If even minor decisions are made high up the command chain and slowly dribble down to the tactical level, there can be no flexible and above all adaptive maneuvers and time is squandered waiting for orders. For Simpkin (1985, p. 53) the promise of maneuver warfare could only be exploited to the full by employing *Auftragstaktik* as a method of leadership.

> If an army cultivates the habit of only doing what is ordered, its movements are bound to be somewhat jerky and intermittent. It will experience an interruption on each occasion of unforeseen circumstances occurring, because all concerned will first await the orders of the higher commanders. (Goltz 1906, p. 226)

Auftragstaktik was designed to answer this need to enable mobility and on-the-spot decision-making. It is impossible to say which came first: the full blooming of *Auftragstaktik* or the new mechanized mobility. Both act as a cause and a result of the other. Mobility led to development of an improved way of command and control and this in turn enabled better exploitation of mobility. To highlight the interconnectedness of mobility and *Auftragstaktik* it has to be stated that tactics, methods, materiel, and technology have to support each other. Svechin (1992, p. 328) noted that different forms of command suit different tactics.

> Positional warfare allows for much greater centralization of command than does maneuver warfare. Hence it is not surprising that as a result of the four-year Sitzkrieg there was a definite trend in favor of command by order, which before the war seemed to be a completely obsolete method of strategic or even operational leadership.

In WW I with its rigid fronts, command could be just as inflexible. Since movements of the troops were minuscule, it was easier to write orders that remained valid for longer periods of time. As we saw earlier, the development from "*Sitzkrieg*" to *Blitzkrieg* was fast and maneuverability couldn't have been effective without proper means of command and control. *Auftragstaktik* made it

> possible, with the help of the new means of warfare, to reacquire the true art of leadership in mobile operations. Individual leadership was fostered on a scale unrivalled in any other army, right down to the most junior N.C.O. or infantryman, and in this lay the secret of our success. (von Manstein 1982, p. 63)

Following *Auftragstaktik* all commanders must be given freedom to decide how to attain their objectives. There are always alternative means to an end and one should choose the one he prefers. To properly employ mechanized troops, Guderian (1956, p. 88) argued, was to give them the ticket to the final destination at once. In the case of attack on France the very last 'railway station' for panzers should have been the coast of the Channel. Even had the chain of command been severed, everyone could have worked independently to reach that goal. The somewhat forgotten lesson of WW II was that the speed of mechanized forces created tactical and operational conditions,

> in which time is fleeting because movements are rapid, command must be far more decentralized than it has been in the past, in order that the actions of subordinate commanders may be immediate. Therefore co-ordination should be sought through general idea rather than through rigid adherence to plan, velocity largely replacing method; but, nevertheless, velocity regulated by a common aim, which is clearly understood by all concerned. (Fuller 1948, p. 55)

After WW II the winners were unashamed to learn from the Germans. Officers of the *Bundeswehr* attempted to explain the meaning of *Auftragstaktik* and lamented that it was translated as "mission type control." This led to confusion because of the Anglophone inability to grasp the subtleties of German. The 'mission' was not a detailed task but closer to what is called 'commander's intent' today. The subordinates were to understand that their mission was to "take immediate action in accordance with the superior commander's thinking' in the absence of a set task" (Simpkin 1985, p. 228). In the absence of clear order there was only a directive of what was to be accomplished. Thus, Simpkin (1985, p. 229) re-translated *Auftragstaktik* as "directive control", arguing that the contrast of directive against an order reflects the spirit of the original. This difference is no longer as clear in established military jargon, since for example Vego (2009, p. IX–11) defined directive as "an overarching term for all the plans, orders, and instructions." The task of the directive is, indeed, to be "the means by which the commander's will or intent is made known to others" (Ibid.) but by definition it encompasses too many different methods of relaying the commander's intent ranging from oral to written orders. The gist of *Auftragstaktik* is that everyone

> must regard his superior's intention as sacrosanct, and make its attainment the underlying purpose of everything he does. He will be given a task of his own, and be told the resources he has to carry it out and any constraints on how he does so. (Simpkin 1985, p. 231)

This argument of the commander's intent being sacrosanct stands as further justification that the 'mission' of the earlier translation is quite descriptive. A subordinate commander is given a task, a time frame to accomplish it, and the resources at his command. These are his constraints and simultaneously his enablers. The most important aspect of *Auftragstaktik* is freedom of thought. It allows the subordinate to act even against the directives, if he is certain his actions support the commander's intent, and still retain the support of his superior. This requires professionalism, since

> directives decentralize commands to a great extent; this will do no harm if there are no centrifugal tendencies in a high command and if there is a general staff which has been indoctrinated to understand the art of war in the same way and is prepared to do battle against parochial interests everywhere. (Svechin 1992, p. 328)

Delegation of authority requires a high level of mutual trust that must be built on respect. Respect only exists among an officer corps that at the same time thinks alike but has full independence of mind and thought with a willingness to harmonize thinking through the entire chain of command. Junior commanders are robbed of the blessing of simple minds to just obey orders. Freedom is more intellectually demanding than detailed orders. As Svechin (1992, p. 327) observed, "command by directives offers great advantages but also poses great dangers if the commanders are unsuitable."

When everyone can act according to his relatively free will for a common purpose, coherence of action is maintained through communication. This took two forms; immediate and full reporting up the chain of command and free, active discussion of the tasks down the chain of command

and horizontally so that commanders could harmonize their activities. Discussion has to span five levels in the chain of command: two levels up and two down (Simpkin 1985, p. 233–239). The only restriction to this communication is time. Commanders should share ideas, intentions, and information, but no extra time is provided for this. If the commander's intent calls for rapid action, time for discussion between levels is drastically shortened. Patton (1947, p. 357) instructed that

> army orders should not exceed a page and a half of typewritten text and it was practice not to issue orders longer than this. Usually they can be done on one page, and the back of the page used for a sketch map. [...] Orders must be issued early enough to permit time to disseminate them. Never tell people how to do things. Tell them what to do and they will surprise you with their ingenuity.

To save time the formulation of orders needs revision. Operational orders today span hundreds of pages and take a long time to compose, but even longer to be assimilated. This lost time could be spent better elsewhere. The ones who execute the orders should have sufficient time for planning their tasks. The lower down the hierarchical ladder one goes, the more detailed the problems and

> much more must be left to the initiative of the individual than in the past. Though the central idea must be maintained, actions should be as flexible as possible. Reports must be as brief as possible and should always, when possible, suggest actions. [...] Time, time, and the saving of it, should be the soul of every order and instruction, of every report and of every message. (Fuller 1943, p. 158–159)

Orders have to be reduced to their bare minimum so that they convey only the necessary message. Once orders, instructions, and communiqués are shortened, time is saved in their formulation, assimilation, and execution. By creating these 'compressed' orders, time can be reallocated where it is urgently needed. The increased demands for effective command and control in contemporary operational art require new inventions in methods. They may not be the ones NCW theorists have propagated, but the principle of unity of command and the ideas and practice of *Auftragstaktik* have to be harmoniously combined. The main requirements of well-functioning command are

> centralized direction and decentralized execution, a high degree of jointness, and interoperability. Centralized direction and decentralized execution are two of the principal prerequisites of successful command and control of one's combat forces. Centralized direction is the key prerequisite for ensuring unity of command. Optimally, it should be applied at all levels of command, from the national-strategic level to the theater and tactical levels. (Vego 2009, p. VIII–8)

The smooth operability of the system of command can be ensured when command is interpreted as direction given in a centralized manner to the subordinates among whom execution is decentralized. Unity of command remains in the hands of the commander but this allows for wider range of tools in execution. Centralized direction is essential for proper coordination of the efforts and objectives of command. It requires making a decision concerning how responsibilities are divided.

> Centralized direction limits to some extent subordinate tactical commanders' freedom of action, but the result is improved command performance. Highly centralized command and control allows the operational commander to make decisions that are better—based on more complete information, obtained from more sources—than those made in a decentralized system. (Vego 2009, p. VIII–8)

Centralized direction, however, doesn't decrease the freedom of subordinate commanders to take independent action if decentralized execution follows the best practices of *Auftragstaktik*. "A military leader cannot hold back in a crisis on the front; a directive should in no way be a form of silence or avoiding responsibility" (Svechin 1992, p. 328). In this system every commander in relation to their own subordinates holds the responsibility and control but does it through guiding directives and his intent, which the subordinates follow. A good commander can delegate authority.

As Strachan (2011, p. 117) noted in the case of the Britain in the World Wars, "the army preached decentralization but practiced control." The natural approach of military leaders is to wish to remain directly in charge of everything and not to delegate authority to subordinates should they make mistakes. This way of thought seems to reside deep in the military subconscious. Vego made propositions intended to make adopting such ideas easier. Unfortunately his logic is occasionally at odds with the principles of *Auftragstaktik*. One example is his argument that "the operational commander possesses a much greater ability to supervise the actions of subordinate tactical commanders, ensuring that their actions are in consonance with his intentions and directives" (Vego 2009, p. VIII–8). If *Auftragstaktik* is properly employed the "ability to supervise" is unnecessary. Officers follow orders, even when they are called 'directives' and 'intent.' It is clear to everyone in the chain of command that if a subordinate knowingly strays from what the commander's intent dictates, he is relieved of command and replaced by a better officer.

To command means more than issuing orders. "Command is a matter for tact and an understanding of human psychology. Some subordinates need and deserve freedom, while others need to be on a tight leash, while still others, who may be great and necessary people, will fly off the handle and have to be persuaded" (Svechin 1992, p. 330). It requires skill to find the limits of what can and must not be ordered to the subordinate based on his character (von Bernhardi 1914, p. 113). One needs suitable people in the chain of command and the chain itself must be flexible. While the 'chain' of command seems to refer to an iron chain which is just as strong as its weakest link, the demand of situational adaptiveness calls for a rubber band. Joint operational warfare theorists often stress flexibility, but seldom elaborate how it can be attained. (Finkel 2011, p. 6)

> Command organization is flexible when it can expand or contract with changing conditions without serious loss of effectiveness. Organizational flexibility is achieved by decentralizing command and control, delegating specific and well-defined functions and responsibilities, and rapidly deploying forces to meet specific situations. (Vego 2009, p. VIII–13)

Authority to make decisions can and should be delegated to trusted subordinates when they are in a better tactical or operational position to decide. Leonhard (1998, p. 201) argued that "the levels of decision making and information flow must be coordinated. That is, the movement of battlefield information should determine who makes the decisions." This is harmonious with the tenets of *Auftragstaktik*. The management of information is a necessity to enable control of Third Wave battlespaces, but the network-centric paradigm of warfare must be capable of directing information flows so that they converge in the operational commander. Information must reach the ones who are expected to make the decisions based on it. Each and every person in the chain of command from the squad leader all the way to the political masters of war have to understand their role in decision-making and resist the temptation of micromanagement. Even if Leonhard (1998, p. 260) claims that "the greater the knowledge of the higher headquarters, the more it can and should effectively employ command," commanders on every level should make only the decisions expected of them and not trespass on the responsibilities of their subordinates.

4.3 Brain and Brawn—Intellect versus Energy in Winning Time

> So, as a prince is forced to know how to act like a beast, he must learn from the fox and the lion; because the lion is defenceless against traps and a fox is defenceless against wolves. (Machiavelli 2003, p. 57)

Machiavelli provided excellent advice what a prince should do, but the same applies today for operational artists who have to embody the virtues of both a fox and a lion while avoiding their vices. Sometimes the intellect delivers when brute force fails. After discussing *Auftragstaktik,* it is suitable to elaborate on the competition in operational art between intellect and imagination on one hand and energy and determination on the other. We can compare the Germans to the Soviets who had a completely different perspective on the flexibility of the chain of command. Tukhachevsky wanted the Red Army to be "strictly brought up in the spirit of firm control, not independent command" (Cited in Simpkin 1987, p. 98). Soviet art of war emphasized carrying out orders unflinchingly. As Leonhard (1991, p. 53) put it, "initiative is neither called for nor permitted. Execution is everything, the hallmark of a great Soviet commander." Both countries were effective with their respective forms of maneuver warfare, but the perspective of command and control was different. While Germany decentralized authority to make decisions, the Soviets centralized it to the extreme. Soviet maneuver warfare didn't depend "upon opportunity and the initiative of subordinates, but rather upon the ability of the commander to impose his will violently and relentlessly upon the enemy regardless of the opposition" (Leonhard 1991, p. 54). We cannot justifiably claim that one method would be more intellectual or energetic than the other. Both approaches to command and control of maneuver warfare require intellect, imagination, energy, and stubborn determination to triumph over the enemy. But they differ in when these qualities exhibit themselves and on which level of command. In the words of Tukhachevsky, "Any suggestion of the exercise of independent command by junior commanders is unacceptable. Not knowing the general situation, junior commanders are always liable to take decisions incompatible with it; and this may engender a catastrophe" (Cited in Simpkin 1987, p. 97). In the Soviet version of detailed control, intellect is employed prior to the battle in painstaking planning of operations. The peak of intellectual activity for the Soviets is

> preparation before the fight begins, as discussed earlier. It is precisely at the same time (i.e. before the fight begins) that they show their imagination, flexibility, and dash. Their system of in-depth reconnaissance, massing of assets, rehearsals, security operations, and deception techniques all reveal energetic, innovative leadership. But after the operation begins, the Soviets are typically married to the plan. After they cross the line of departure, there is great pressure upon the Soviet commanders to conform to the plan. Opportunity recedes into less significance than momentum and tempo. (Leonhard 1991, p. 54–55)

The Germans initiated their operations with short orders and general directives emphasizing the commanders' ability to innovate and energetically search for opportunities to be exploited. In other words, there are two places where one can win time in command. One is in planning and the other is in execution. Correspondingly there seem to be two types of soldiers with their characteristic approaches. One uses the intellect, like von Moltke, Suvorov, or Napoleon, and the other inexhaustible energy like Patton, Rommel, or Guderian.

von Seeckt (1930, p. 103) argued that at the heart of warfare is the concept of action. It is divisible into three stages: the decision to act which had been given birth by thought and analysis, the preparation, and the execution of the chosen action. In all stages action is subordinate to the will. The will is born out of character and it is in turn more crucial than the spirit. The spirit without the will is worthless and the will without the spirit outright dangerous. To win time requires people who can combine will, spirit, intellect, thought, analysis, and energetic execution into one. Operational art doesn't need either thinkers or doers but people capable of both and acting without loss of time. This need to plan with perseverance, intellect, and imagination and then to execute the plan with will, energy, and determination means that

> two very different things must exist in a man to make him a general: he must know how to arrange a good plan of operations, and how to carry it to a successful termination. The first of these talents may be a natural gift, but it may also be acquired and developed by study. The second depends more on individual character, is rather a personal attribute, and cannot be created by study, although it may be improved. (de Jomini 2007, p. 247)

von Clausewitz (1989, p. 229) argued that "among all the military virtues, the energetic conduct of war has always contributed the most to glory and success." There is an uneasy relationship between the thinkers and doers or the intellectual and energetic types. Commanders tend to favor one and discard the other based on their personal traits, but both are essential. Meticulous study and rigorous training are required to perfect operational art in both the mental and the physical sphere of activity, but they don't always coincide in the same person. Patton was an educated and a well-read man, but his knowledge of the theories of war from distant past to present didn't deprive him of his reliance on bold methods and direct means. As he wrote, "I have never given a damn what the enemy was going to do or where he was. What I have known is what I intended to do and then have done it. By acting in this manner, I have always gotten to the place he expected me to come about three days before he got there" (Cited in Carver 1979, p. 82–83). Patton's energy and willpower in pushing his troops onward relentlessly enabled him to win a temporal superiority over the enemy and master time in his operations.

As Rommel (1953, p. 119) saw it, "a commander's drive and energy often count for more than his intellectual powers—a fact that is not generally understood by academic soldiers, although for a practical man it is self-evident." The energetic approach has its benefits. The academic soldier may through his meticulous planning save time by preparing a surprise that enables a quick victory. In the heat of the combat, however, anything that requires preparation has to be cast away and the energy and will of the commander triumph. Energetic commanders drive their troops onward and maintain momentum. Rommel characteristically focused on the tactical requirements. On the strategic level the intellectual powers of commanders dominate over their drive and energy. von Moltke was an exemplary specimen. But an operational artist must combine both to manage time. On the operational level and tactics, Rommel serves as an example of maneuver and time management. He was an intelligent, but also a practical and dynamic, commander.

> One of the first lessons I had drawn from my experience of motorized warfare was that speed of manoeuvre in operations and quick reaction in command are decisive. Troops must be able to carry out operations at top speed and in complete coordination. To be satisfied with norms is fatal. One must constantly demand and strive for maximum performance, for the side which makes the greater effort is the faster—and the faster wins the battle. (Rommel 1953, p. 225)

Such an energetic approach is harsh for the troops and the commander alike. Speed of maneuver and quick reaction in leadership are important, but striving for maximum performance requires careful planning of how to set the tempo and rhythm of operations. Patton wrote concerning how long a continuous push can last before rest that, "infantry troops can attack continuously for sixty hours. Frequently much time and suffering are saved if they will do so. Beyond sixty hours, it is rather a waste of time, as the men become too fatigued from lack of sleep" (Patton 1947, p. 353). The commander must know where the point of maximum bearable strain of his troops is and stop them just shy of reaching it. In tactics one wins time by squeezing every drop of energy out of his troops, but operationally actions need to be synchronized so that a fresh unit can take responsibility of upholding the tempo and momentum when energy of the first one is spent.

However, even Rommel was not all about energy and drive. He understood that maneuver warfare and mechanized mobility increased unpredictability. There were too many options and different courses of action open to make any binding plans. Rigidity of mind and plan had to give way for flexibility and a reactive stance. Battlespace is the playground of the unexpected and

> because of the great variety of tactical possibilities which motorization offers it will in future be impossible to make more than a rough forecast of the course of battle. This being so, the issue will be decided by flexibility of mind, eager acceptance of responsibility, a fitting mixture of caution and audacity, and the greater control over the fighting troops. (Rommel 1953, p. 517)

Even if the mind had to be flexible to master operational art with its sweeping maneuvers, the commanders of the mechanized age had to be strict and unyielding. "Mental conception must be followed by immediate execution. This is a matter of energy and initiative. What the soldier needs is a combination of realist intellect and energy. Whatever is attempted must be carried through" (Rommel 1953, p. 518). Rommel boasted about his ability to plan quickly and in a short time create simple but imaginative plans. As he wrote to his wife from Africa,

> I'm full of plans that I daren't say anything about round here. They'd think me crazy. But I'm not; I simply see a bit farther than they do. But you know me. I work out my plans early each morning, and how often, during the past year and in France, have they been put into effect within a matter of hours? That's how it should be and is going to be, in the future. (Rommel 1953, p. 179)

The commander must possess the gift of imagination, picturing in his mind several alternative future courses of development and choosing the most likely. This is what Rommel meant by 'seeing farther.' Imagination and forethought were used to draft the outline of an operation to be executed with energy. In the words of Liddell Hart (1948, p. 61), Rommel "had a real touch of genius in the tactical field, combined with dynamic power. He had a flair for the vital spot and the critical moment." Rommel had a profound sense of how to relate time to space and to dynamically create spatio-temporal force concentrations. Nevertheless, his ability to grasp the grand scheme of strategy proved insufficient in the longer time-perspective. He admittedly focused on the

> power of execution, the ability to direct one's whole energies towards the fulfillment of a particular task. The officer of purely intellectual attainments is usually only fitted for work as an assistant on the staff; he can criticise and provide the material for discussion. But a conclusion intellectually arrived at needs the executive power of the commander to follow it up and force it to realization. (Rommel 1953, p. 288)

Time is consumed in different things on different levels of warfare. The grand strategist doesn't worry about execution and even strategists do so to only a limited degree. Their task is not to look at a watch but at a calendar. They concentrate on the intellectual side of the art of war where demands are different than in tactics. Rommel, like many other panzer-commanders, were the men of action, willing to lead from the front, that mechanized operations demanded (Hart 2006, p. 10; Stein 2007, p. 35). The strategist, the operational artist, and the tactician need each other for the fulfillment of their respective tasks. Energy and rapid action are important in tactics while understanding the flow of time as a combination of different currents or waves, and exploiting them is a strategist's task. These somewhat contradictory qualities should be well-balanced in the operational artist. Military profession values strong commanders with iron will and determination but they may experience complications due to fog and friction of the battlespace. Strong commanders don't always perceive the need for change when the operation doesn't proceed as planned. They attempt to use their will to make the reality conform to the plan. "People are very obstinate, but among the most obstinate are great military leaders" (Svechin 1992, p. 330).

Choosing the safest roads after careful analysis seldom leads to victories, but unthinking action is hazardous. Endless contemplation may provide a perfect plan but too late. As Brian Holden Reid (1998, p. 2) wrote, "soldiers need brawn as well as brains" and the commander should favor one over the other depending on the situation. Seizing the initiative should be the crux of every operational plan. To seize it one has to have the permission to act as he sees best. The importance of initiative is elevated on operational and strategic levels. Strategy and operational art should be based on "action adapted to circumstances, and, consequently, concentration in strategy may be defined as making the most of opportunity and also of forecasting and foreseeing the possibility of opportunity before it arises" (Fuller 1923, p. 34). The operational commander requires intellect to evaluate the current moment and make causal predictions of what will occur. The ability to deduct and estimate future enemy actions and countering them before they have actualized determine his success. The energy of the tactical commander is as valuable for winning time as the intelligence of the strategic commander. Next we need to look closer at the operational commander to discover what kind of intellectual support raw energy requires to manage time and set the rhythm of operations.

4.4 Balancing Imagination and Intellect to Manage Time

> For Third Wave civilization, the most basic raw material of all—and one that can never be exhausted—is information, including imagination. (Toffler 1990a, p. 351)

In indust-real warfare the most important material for the war effort was either factory-produced weapons or masses of men in uniform. In Third

Wave warfare information allegedly is the most important asset, and as Toffler wrote, imagination is part of information and its management. Emphasizing imaginative qualities adds artistic element to warfare but does not necessarily reduce the need for scientific approach to it. While certain writers of the Second Wave, like Fuller and Triandafillov or de Jomini before them, have attempted to turn the art of war into science, many still consider it an art or a combination of both. The art of war deserves the same treatment as other arts. There are those who create the masterpieces, those who criticize and evaluate them, and those who make art their lifetime study. For those who wish to understand art some eccentric ways of the artists are frustrating. Vincent van Gogh produced masterpieces but also mutilated himself. Salvador Dali was not exactly a picture of sanity. Similarly some military men are original thinkers. Among the mass of military men that abhors the esoteric intellect of Fuller, Montgomery is highly respected. He wrote how

> the C.-in-C. of great armies in the field must have an inner conviction which, though founded closely on reason, transcends reason. It is this which will enable him at a certain moment in the battle—the right moment—to take a short cut which will take him to his objective more swiftly and more surely than equally careful but less inspired commanders. (Montgomery 1961, p. 51)

Supposedly inner conviction transcends reason and gives an edge to the less inspired. For Montgomery, first everything is founded on reason and then reason is discarded because the inner conviction, be it religious or something else, surpasses it. This suggests that a successful commander and operational artist must transgress the boundaries of reason. To quote Patton (1947, p. 8), "in forty hours I shall be in battle, with little information, and on the spur of the moment will have to make most momentous decisions, but I believe that one's spirit enlarges with responsibility and that, with God's help, I shall make them and make them right." Conviction may help the commander to cope with the pressure of momentous decisions. Even in the twenty-first century secularized western world some commanders rely on their faith for inspiration or a source of strength (Franks 2004, p. 430). This type of inspiration is external and fickle just like *fortuna*.

As a Satanist, Fuller sought from within the personality of a man ways to surpass what could be expected of him. The commander must perceive the status quo, estimate what will happen, and make decisions in accordance with his objectives even when he is ill-informed. The applicability of these decisions for the circumstances and the quickness of reaching them defines generalship. Fulfilling the tasks that require almost a superhuman personality from the commander demands what Fuller called genius. This baffling and almost inexplicable concept elevates generals into great captains. Fuller was intrigued by the idea of genius and struggled to describe how it

> is neither high talent, nor outstanding intelligence, nor is it a product of learning, or of discipline or training. It is, so it would seem, a creative gift, intuitive and spontaneous in its manifestations, that endows its possessor with a god-like power to achieve end which reason can seldom fathom. (Fuller 1960, p. 82)

Behind the façade of Fuller's fancy for grandiose expressions such as "god-like," we can glimpse crucial elements of this elusive concept. Genius is not related to extraordinary powers of intellect but rather a certain mindset and a way of thought. Creative, intuitive, and spontaneous: these three are characteristics of a mind that can aspire to genius. It is about achieving things through intuition and imagination that exceeds the limits of where reason can take the commander. A mediocre man can handle routine duties, but sudden improvisation, being prepared for all eventualities in all situations, and making the right decisions presuppose a man with exceptional mental alertness and understanding of temporality. He must manage time but also know "how to profit from that favorable moment which occurs in all battles and which decides their success" (de Saxe 1944, p. 118). Operational artists require imagination to achieve greatness. The military theorists are, however, divided on this. The unimaginative ones emphasize rational calculations. For Foch,

> in strategy, as in any other business, a leap in the dark is the reverse of sound action; nobody has the right to substitute for actual facts, which must always be sought, the productions of imagination, mere suppositions. On facts alone can a rational manoeuvre be founded. (Foch 1920, p. 250)

Foch is both right and wrong, as he often was. Indeed, all military activity should be based on rational grounds, but this creates just a practicable course of action. Imagination paired with intellect allow the commander to aspire for operational surprises. When the tactical commander makes his decisions in the heat of battle, they must be based on facts and not his fancy. In operational art facts erect the foundation for imaginative ideas which should be used to produce something original especially when time is in short supply. One must keep searching for the *kairos*-moment. To seize the moment requires what Fuller had in mind when he described genius as

> one of those apparently inexplicable powers which differentiates the truly great man from the normal. It is not an instinct, for otherwise it would be common property; it is not the reason, as we usually understand it; but as it accomplishes in an incredibly short time a purpose which the faculty of reason would attain by a slow and no more certain progress, it, I think, may be considered as the highest dimension of this faculty. (Fuller 1926, p. 98)

A genius could take the "short cuts" of Montgomery. Genius is not about instinct, since while instinctive action saves time, acting on instinct equals acting on one's emotions. Genius combines instinct and intelligence to informed action on the spur of the moment. To possess genius is to manage time without being tied to the decision-making cycles and so distort the enemy's rhythm.

> What the man of genius does is to imagine automatically, and so produce original relationships which, metaphorically, are born patented, since others can seldom copy them. If I may hazard to set down the qualifications of the great captain, then I should say that they are:
>
> 1. Imagination operating through reason.
> 2. Reason operating through audacity.
> 3. And audacity operating through rapidity of movement.
>
> The first creates unsuspected forms of thought; the second establishes original forms of action; and the third impels the human means at the disposal of the commander to accomplish his purpose with the force and rapidity of a thunderbolt. (Fuller 1926, p. 100)

Fuller's formula illustrates how in operational art imagination manifests itself through reason. We tend to treat these as binary concepts and differentiate between imaginative and logical thought. Combining these two creates audacity for the actions of the commander, and audacity combined with rapidity produces results. Imagination is a prerequisite of originality, and original ways of addressing problems create unpredictability.

> To copy is not to originate, and originality of thought is the mental co-efficient of the principle of surprise, and, when determination to win is accentuated by this principle, frequently an objective can be created by one side which is totally unrealized by the other. Such a creation is what I call tactical forethought—seeing an action before it is fought. Foresight is the fruit of the scientific method, and it must not be confounded with imagination. Imagination presents to us a possibility, reason analyses it and stamps it with a value; these two are the parents of foresight, which is nothing more than mentally standing on tiptoe. (Fuller 1926, p. 281)

Operational artist wins time through his ability to "foresee" the future. However, operational foresight and forethought cannot be achieved by an imaginative mind alone. It requires reasoning, deduction, and analysis. Imagination is indispensable but in determining practicable courses of action. The operational artist must

> foresee the situation as it will be at the expiration of two, three, or even more days. Jomini extols this quality in Napoleon, and attributes to it the rapidity and ease of all his arrangements. The positions of his corps, division, and brigades at any given moment were always present to his mind. He therefore forgot nothing, and never failed to notice chance means to the end in view; he thought of things which everyone else would have forgotten, and was rich in inspirations. Such is principally the work of imagination. (Goltz 1906, p. 36–37)

Fog of war makes it impossible to calculate how events will proceed over a time-span of even a few days. The operational artist, like Napoleon in his time, must visualize not only the likely and unlikely outcomes, but the intricacies of the moment itself. Lacking imagination, the troops remain only arrows on a map or a computer screen, but they become 'alive' in the mind of the imaginative commander. Foresight of possible courses of development wins time, because the operational artist reacts faster to any eventuality, since he has already determined his response. Napoleon gave credit for his preparedness to imagination, but he spoke of meditation.

> If I appear to be always ready to reply to everything, it is because before undertaking anything I have meditated for a long time—I have foreseen what might happen. It is not a spirit which suddenly reveals to me what I have to say or do in a circumstance unexpected by others; it is reflection, meditation. (Cited in Fuller 1926, p. 42)

Imagination as free play of thought offers the general a chance to invent something unexpected, but unrestricted imagination is hazardous since it produces fantasies disconnected from reality. "The power of imagination must know how to conjure up a picture that shall serve us as a basis for subsequent action; and in order to prevent distortions and dislocations, the commander must not fail to exercise his fancy and to keep it working" (Goltz 1906, p. 37). Imagination is a two-edged sword that needs to be wielded cautiously. The best way to save time with imagination is to allow it free play to augur further developments in the course of operations, but operations must be planned and executed with somber realism.

Patton was seldom blamed for being overtly imaginative. Despite his realist stance toward war as a phenomenon and a robust boastfulness in his ways, he set aside a role for imagination. He wrote once awaking in the morning with a complete attack plan for two corps in his mind. "Whether these tactical thoughts of mine are the result of inspiration or insomnia, I have never been able to determine, but nearly every tactical idea I have ever had has come to my head full-born, much after the manner of Minerva from the head of Jupiter" (Patton 1947, p. 238). Patton was a great general and a braggart of comparable scale and this ability to create detailed plans offhand is an exaggeration. The capability to imagine allowed Patton to draft his plans rapidly and resulted in huge timesavings. Another operational artist, von Manstein (1982, p. 105), wrote that

> naturally I did not immediately find myself presented with a cut-and-dried operation plan in that October of 1939. Hard work and endeavour must always confront the ordinary mortal before he attains his goal. No ready-made works of art can spring from his brain as did Pallas Athene from the head of Zeus.

Both men employed considerable amounts of imagination to produce efficient plans. While history cherishes its Napoleons as imaginative artistes there are famous butchers who waded to victory through blood by the size of their armies. Generations of military theorists lamented how soldiers don't take pains to study war. Military intellectuals, defined as thinking, serving soldiers, interested in ideas and imbued with the ethos of the profession of arms, are still too rare (Reid 1998, p. 3).

> Science is by definition worth studying. But when it comes to art, there are those who study it and they more often become critics than artists. What the true artiste needs, is inspiration and vision and from these sources springs the perspiration that creates the actual masterpiece—or spoils the canvas. Art and science are not commonly practiced by the same person and indeed, the psychological composition of a person who excels in one, commonly fails in the other. If we want our commanders to be scientists, we are likely to make no mistakes and use our forces in a manner dictated by logic and reason. If we want them to be artistes, we need to accept the fact that many failures are often needed as one produces the sketches of his future masterpiece. The great work of military art may be preceded but by shedding a huge quantity of blood. (Fuller 1926, p. 20)

Highly populated countries may allow their artistes to be wasteful in the creation of their blood-soaked masterpieces of operational art. Small nations don't have the luxury of squandering lives and time in training their generals. No sketches can be scribbled. It is thus natural to turn to educating the commander to be scientific. Simultaneously they become highly predictable administrators of war who may not suffer devastating losses, but neither will gain immense victories. As Lawrence (1935, p. 193) argued, "nine tenths of tactics were certain enough to be teachable in schools; but the irrational tenth was like a kingfisher flashing across the pool, and in it lay the test of generals." Inspiration and vision cause surprises because their realm is the unexpected and the spur of the moment, but the risks are enormous.

A scientific soldier can manage quantifiable time better. He can trim his military machine so that not a second is lost. He is an exemplary specimen of the indust-reality. The artiste, however, relying on his inspiration and his grasp of the rhythm of battle, can potentially work on a different timing to break the harmony of the enemy. One can either treat time as a quantifiable resource measured in seconds and minutes, or as a flow of *kronos*-time into which the operational artiste dives kingfisher-like at the precisely right *kairos*-moment to subdue his enemy. Third Wave warfare requires a combination of the two. It is very seldom that qualities of measuring scientific time and auguring the perfect moment conjoin in the same operational commander. Indeed,

> armies are not composed of philosophers, either at the top or at the bottom. In no activity is optimism so necessary to success, for it deals so largely with the unknown—even unto death. The margin that separates optimism from blind folly is narrow. (Liddell Hart 1936, p. 69)

How to cultivate a genius suitable for operational art of Third Wave warfare? For Fuller as an indust-real theorist the answer was by creating a scientific method for understanding and disseminating the essence of warfare. But no matter how much time and energy is spent in training future commanders to be geniuses, the desired results are unlikely. "The first master of the art of war is experience, the second is reason, and the third, and greatest, is genius. Experience can be bought at its price; reason can be obtained by study and by reflection; but genius would appear to be God's gift" (Fuller 1926, p. 99). To combine artistry of war with rationality demands that a successful commander possesses a practical outlook. Nothing is free and while

> intellectual power makes up a large part of what we term "great military genius," and for this reason the officer who is training for high position in war should endeavour to develop his reasoning powers. But he must do this through constant, critical examination of the past and present, rather than through forms of metaphysical speculation. (von Freytag-Loringhoven 1991, p. 184)

Metaphysics of war is not a highly recommended realm of interest in an officer. It is commonly thought that he should focus on the accumulated military knowledge and think logically. Visionaries tend not to be great organizers and executors, but the organizers seldom create anything original. The right combination is somewhere in between so that the two balance each other. Intelligence doesn't equal genius without intuition.

> An inner understanding, his own intuition, usually provides the only means for penetrating the fog of war and controlling events properly. The instinctive feeling of the proper action to take at the moment has always been present in every great leader. (von Freytag-Loringhoven 1991, p. 262)

Genius can be enhanced by experience and training, but if it isn't an inborn personality trait, it can't be created out of thin air. Fuller admired Alexander as a military genius par excellence and praised his spontaneity. What made him a world-conqueror was

> the startling rapidity with which he always acted: no situation caused him to pause; all difficulties were immediately stormed; though risks were immense, to him success seemed foreordained. Time was his constant ally; he capitalized every movement, never pondered on it, and thereby achieved his end before others had settled in their means. (Fuller 1960, p. 82)

This illustrates why we have discussed genius when the focus is on saving time in warfare. Spontaneity, intuition, and intellect together supplement the energetic action of a genius and enable him to rob time from enemy and save time for his troops by capitalizing every moment. Time is an ally of the genius because of his ability to act at the right *kairos*-time while both his mind and plan remain agile and flexible in execution. The admittedly rare military genius can

> produce original combinations out of the forces of war; he is the man who can take all these forces and so attune them to the conditions which confront him that he can produce startling and, frequently, incomprehensible results. (Fuller 1926, 98–99)

4.5 Original and Rapid or Traditional and Slow

Thucydides wrote that "it is just as true in politics as it is in any art or craft: new methods must drive out the old ones. When a city can live in peace and quiet, no doubt the old-established ways are best: but when one is constantly being faced by new problems, one has also to be capable of approaching them in an original way" (Thucydides 1971, p. 52). On the level of operational art there is no peace and quiet, but constant motion problems galore. There is a dichotomy between the challenges command in war places upon the officer and his training. Liddell Hart praised the Napoleonic system of 'read and re-read.' To be creative, the library of military history "is still the only sure foundation for command in war, and study and reflection are the almost essential complement to the natural gifts of leadership—will and originality" (1927, p. 195). For Liddell Hart (1932, p. 323) originality was the most vital of all military virtues. Studious learning can provide an aspiring operational artist the keys to the treasure trove of originality, and to be unorthodox he must be familiar with the orthodox. He needs to know the dogma and canon in order to breach them. Studying the classical methods enables him to adhere to the same basic principles as the masters of the past but elaborate on them and draft original operational art.

"All artists, and particularly those in the military sphere, derive much benefit in their training from studying the masters and masterpieces, for there is something contagious about magnificence. Yet this conformity must neither become exclusive, nor this imitation servile" (De Gaulle 1976, p. 125). Magnificence of operational art unfortunately isn't contagious. People striving for it fail to differentiate between inspiration and perspiration. Yet, as Turpin declared, both study and genius are required since "nothing but a mind enlightened by a diligent study can make a due application of rules to circumstances" (Cited in Gat 2001, p. 38). The key to originality can be found by study of a wide range of different topics and subjects. Especially in asymmetric operational art the commander's viewpoint and understanding depend upon the extent of his

> contemplation of matters other than the swords. It has been posited that creativity is essentially the ability to mentally connect two ideas that were previously unconnected. If so, then the potential for creative military thought lies in the officer who cultivates an appreciation for non-military things. The ability to imagine an innovative solution to problems in war rests not only upon military competence, but also upon a wide interest in other areas of life. (Leonhard 1998, p. 238–239)

The requirement of innovation and originality is at odds with inoculation of a young officer into his profession. It is a challenge to adopt and adapt existing ideas, traditions, and practices and to utilize them in an original way. Officers should remain open to unorthodox ideas throughout their careers. As Wavell instructed younger officers, "never let yourself be trammelled by the bonds of orthodoxy; always think for yourself; get as much experience outside the ordinary run as you possibly can; and remember that the herd is usually wrong" (Cited in Ferguson 1961, p. 41). But unorthodoxy and running independently of the herd is not the quickest career path to high command. Therefore,

> miraculously blessed is he whose mental horizon, when he has reached a position of command, remains unclouded by the traditions, social and professional, he has imbibed; whose experience has not been bought at the price of some measure of his imagination, mental independence, and receptiveness to new ideas. (Liddell Hart 1927, p. 177)

The demands for originality and abiding by the traditions, procedures, and accepted forms of military action are contradictory. Military education has to teach a unified way of acting and decision-making to create an army instead of isolated individual thinkers. This provides the armed forces with efficient staffs and tactical commanders. Out of this mass of officers only a few talented ones rise to the highest ranks. They need to be original thinkers even though they have followed the same military education path as others. "No system of Staff College training, however far developed, can escape the danger, because of its very nature, that it may become a factory for the mass production of stereotyped brains" (Liddell Hart 1927, p. 182). The brains guiding operations need to be prototypes to surprise the enemy. As Liddell Hart (1927, p. 174) wrote, "surprise is the supreme virtue of warfare, originality of mind the quality which breeds it." Unless the commander possesses an original way of thought he is incapable of true surprises that rob the enemy time.

Practice makes perfect, but perfection is an enemy of good in tactics and operational art which are more time-sensitive than strategy. This is because "long training tends to make a man more expert in execution, but such expertness is apt to be gained at the expense of fertility of ideas, originality, and elasticity. War is the realm of the unexpected; and adaptation to the unexpected comes more naturally to youth" (Liddell Hart 1950, p. 314). In this and many other things, Fuller saw eye to eye with him. For Fuller war was a young man's business (Reid 1997, p. 125; Fuller 1933). Originality and elasticity diminish with age and professional experience. A mind absolutely ignorant of military practices might formulate a truly unorthodox plan. However, originality must conjoin with practicability and some ideas may not be executable. Often those best suited for command are not the ones with original ideas and some of the extremely original thinkers must be kept far from command positions. Commanders stick to their experience, but they have staffs to produce fresh ideas for them. Here is a role for a younger, fresher mind. It can incorporate original ideas into the plan and ultimately the unbending will and energy of the commander executes them.

Again using Napoleon as an example, the man who scored great victories early in his career through audacity, willpower, and energy and who after a long march rushed upon the enemy continuously rallying his generals with "*activité, activité, vitessé*" was not the same man who was defeated in Waterloo. In Ceva, for example, Napoleon's youth provided a relative advantage on his enemy (Colby 1943, p. 114–116). Later his age became a hindrance. Napoleon himself admitted that

> 'one ages rapidly on battlefields.' And at the time he said it he was in the second year of his career as a general. Since then in the course of 17 years, many things had happened, bound to shake even this giant. The accumulation of many sins was gnawing at the marrow of this Titan. Halting or turning back was impossible. He was driven forward, ever forward, against ever-increasing obstacles. To oppose them he lacked the strength. His fall was imminent, if not on 18 June, then later. It was inevitable. (Schlieffen 1936, p. 59–60)

Napoleon lost his mental youthfulness in combat. Alexander was also a young man when he conquered the known world. In the words of Quintus (1988, p. 36), "his age gave added luster to all his achievements for, though hardly old enough for undertakings of such magnitude, he was well up to them." Alexander died young and avoided the fate of Napoleon. More than anything, the flow of time led to Napoleon's final demise. The energy of young age is quickly spent in war. The freshness of novel ideas is temporary and with time these ideas ripen and rot. To retain the capability to surprise and reinvent his methods, an operational artist has to recreate himself time after time and not allow his ideas to solidify. This is impossible to repeat endlessly and seventeen years drained Napoleon. The same is bound to happen to any general forced to fight throughout his career. Through inevitable aging, a commander will become a casualty of time.

We should not, however, designate all older military men as mental dinosaurs of a bygone era. Throughout history there have been late bloomers among the great captains. For example Suvorov developed with age. His 1793 war plan against the Turks shows he still had

> daring thoughts, and refutes those who claim that he was no planner but a man who acted impulsively without ever reckoning the complex factors that influence war. In its freshness and economy it confirms that, like Turenne, Suvorov was one of the few generals who actually improved with age. (Longworth 1965, p. 184)

It is not about physical age but mental youth that manifests itself as originality and flexibility of thought. Some people are born reactionaries and conservatives while others keep an open and agile mind in their advanced years. Even if Suvorov, Marlborough, Cromwell, or Patton can be evoked as examples of the suitability of experienced professionals, there are agile minds like Napoleon or Frederick the Great who aged badly on the battlefield (Wavell 1941). As Frederick phrased it "my last days are poisoned and the evening of my life is as horrible as its morning" (Cited in Bond 1998, p. 27). This problem with old commanders is heightened on one hand by the increased pace of warfare and the highly advanced technology on the other. Already in mechanized warfare "the most important need that remains to be faced is a reduction in the age of general officers, which is dangerously high in view of the quickened tempo of modern operations and their increased strain on mental as well as on physical resilience" (Liddell Hart 1932, p. 255).

While many officers of the older cadre may retain top physical form, much of the strain in the Third Wave battlespace is mental and psychological. Tempo is problematic. Advances in technology have enabled faster operations and creation of desired effects in fractions of the time that was spent even some decades ago. Some indust-real generals were unable to control the new surge in mobility mechanization created and this provided more agile minds like Guderian or Rommel with an edge. Rotary wing movement, airborne troops, and automated or even autonomous weapons systems have further accelerated that speed. As of today, it is still unclear just what kind of compression of time in operational art may occur during the high tide of the Third Wave.

de Jomini (2007, p. 40) asked a question that is as relevant today as in his time. "Those who have served long in peace will be at the head of their arms or corps, and will have the rank appropriate for this position; but will they always be the most capable of filling it?" Seniority of age and rank doesn't automatically provide the ability to command operations. Seniority may as well equal senility. One cannot have only youngsters in high command but operational artists need to be imbued with mental elasticity characteristic of youth and antagonism to established procedures. While Foch praised the audacity of young commanders, he argued that because of youth and resulting mental immaturity they had to be "commanded and guided" since "they lacked experience which alone can develop judgment, and the habit of authority, which alone can ensure to an officer the calm confidence which leads to wise as well as vigorous decisions" (Foch 1931, p. lxi). This will intellectually ossify the young officers. The answer may be continuous training of the commanders by putting them into unfamiliar situations and forcing them to make decisions that contradict the prevailing dogma. To remain original, the operational artists should not be allowed to perfect their methods to supposedly fit each and every situation, but force them instead to perfect their decision-making process so that they could function in the most varying conditions. The gateway to originality is in flexibility of both the commander's mind and plan.

In operational art one must grasp the implications of the accelerated pace and flexibly adapt. Toffler (1990b, p. 344) argued that when a person, in this case the commander, is forced to "operate above his adaptive range" something called "future shock" will emerge. It is comparable to a culture shock that occurs when a person suddenly gets immersed into a culture that is fundamentally alien and incomprehensible to him. Without mental preparation to the effects of compression of time and acceleration of pace, future shock is almost unavoidable. Accelerated pace cannot be compensated by being 'mentally quicker' or just by doing same things 'more effectively'. Future shock is "an overload of the human organism's physical adaptive systems and its decision-making processes. Put more simply, future shock is the human response to overstimulation" (Toffler 1990b, p. 326). In the chaotic conditions of the battle and the operation soldiers and commanders operate in the upper reaches of their adaptive range and occasionally are pushed beyond their limits (Toffler 1990b, p. 344). As the tempo of the battle changes anew in Third Wave warfare, one must readjust one's inner rhythm. No one is more adept to survive future shock because of accustomization to similar effects. Rather the limits may emerge sooner. Future shock is a reality for everyone but some naturally adapt better than others. In addition, the input overload affects worst the commanders who have to make decisions in the compressed time; They have

> to adapt to a new life pace, to confront novel situations and master them in ever shorter intervals. We are forcing them to choose among fast-multiplying options. We are, in other words, forcing them to process information at a far more rapid pace than was necessary in slowly-evolving societies. There can be little doubt that we are subjecting at least some of them to cognitive overstimulation. (Toffler 1990b, p. 355)

War as an art form fits into the old adage of *ars longa, vita brevis*. A complete comprehension and mastery warfare in all its forms is impossible to attain. "It is sad to remember that, when anyone has fairly mastered the art of command, the necessity for that art usually expires—either through the termination of the war or through the advantaged age of the commander"

(Patton 1947, p. 366). It is only during warfare that commanders can experiment and develop their skills in tactics and hope for a chance to become operational artists. As Strachan (2013, p. 210) wrote, "it is in the exercise of operational art that today's senior generals (…) hope to reach the acme of their professional careers." Their time in high command positions is short. They can write their memoirs so that consecutive generations and their aspiring operational artists can peruse on these thoughts and experiences and not build their art from scratch. Because

> knowledge of the art and science of war is not necessarily conferred by years of routine soldiering, that the book of military knowledge is open to all, and that those few who have mastered its contents by intensive study may include representatives both of the professional and also of the amateur soldier. (Liddell Hart 1927, p. 175)

Just as in any other artform, it is not always a necessity to attend an art school. If the passion to learn, study, and create burns bright, the art of war can be mastered by a civilian as well as a soldier. Liddell Hart proves this point. Despite his experience in WW I and resulting rank of captain he never was a professional soldier. As Danchev (1998, p. 31) wrote, "it was the art, perhaps the artifice, and not the article that fascinated him." Despite his artistic leanings, or perhaps because of them, Liddell Hart profoundly affected the development of operational art. One might benefit from being unaccustomed to military way of thought to create something unorthodox. With a quick glance of military history

> we find that most new ideas, which eventually materialize into theories or concrete form, originate in piratical exploits outside the existing military organization, and that only after a period of virulent abuse do they become adjuncts or undesirable foster-children in the military family. (Fuller 1923, p. 73)

With youths or with experts from non-military background the armed forces can learn new ways of addressing old problems. These minds are not cluttered with dogma but free to roam and invent new and unorthodox measures and to carry them out with boldness uncharacteristic to accumulated age and experience. Antonio Gramsci wrote that most of the original thinkers of the world are institutionalized. While he meant something completely different, the military is an institution in need of original ideas supplied by a

> 'creative mind'; because to term it 'inventive faculty' appears to us too shallow. There are but few men who have original thoughts. Ben Akiba's saying 'Nothing new under the sun!' is as true of the world of ideas as that of phenomena. Most people in these days only make use of what they have inherited or acquired. But in war situations are of such nature that they appear similar without ever being quite the same. The number of causes and forces is too great to admit of absolute similarity. The general cannot, accordingly, employ the exact means that have been already adopted on a previous occasion. At any rate, there will be something entirely new in the manner of their application. Some slight addition of personal invention is always necessary, and that requires the aid of the ever productive power of the creative mind, as well as the will to employ it. (Goltz 1906, p. 38)

An operational artist must combine old ideas in novel ways and add a touch of originality. This requires creativity and nonconformity. As Strachan (2011, p. 98) argued one of the distinctions "between operational art and doctrine: the former can be an individual matter, whereas the latter is collective." To rise above the doctrine is not for the stereotyped brain, because at the moment he does it, a gulf opens between him and the established collective perception of the outlook of warfare. Yet, for the gifted, it may just be the breach needed to gain an advantage over his enemy and a move that turns mechanical conduct of operations into operational art.

> The danger of a doctrine per se is that it is apt to ossify into a dogma and to be seized upon by mental emasculates who lack the power of analytic criticism and synthetic thought, and who are only too grateful to rest assured that their actions, however inept, find justification in a book which, if they think at all, is, in their opinion, written in order to exonerate them from doing so. In the past, many armies have been destroyed by internal discord, and some have been destroyed by the weapons of their antagonists, but the majority have perished through adhering to dogmas springing from their past successes, that is self-destruction, or suicide, through inertia of mind. (Fuller 1923, p. 33–34)

4.6 Flexibility of Mind and Plan

> And so we come to the plan—which is not a detailed programme of specific steps matched with equally specific tolls, or in the case of the military, materiel and units. Rather, in the new approach the plan should be a broad outline, an intended pattern of events, based on the information and analysis to achieve the desired outcome, enumerating the objectives to be achieved; and allocating responsibility, authority and resources accordingly—so that effects achieved are coherent, focused and networked. (Smith 2008, p. 391)

As Smith argues, a plan in contemporary warfare needs to resemble a sketch which the artist further elaborates. It is a framework that unites efforts and effects but allows for improvisation as long as the desired outcome is reached. A common mistake in operational planning is to create a 'perfect' plan laboring under the illusion that is could somehow lead to perfect execution. "Clever men usually look too far afield for the best method, and fail to perceive the paramount importance of the timely adoption of a practical method" (Goltz 1906, p. 31). This is the "mechanical" aspect of warfare Wavell (1953, p. 41–42) wrote about, but the most detailed plan is useless unless it can be finalized in time. A shoddy but timely executable plan is always better than an intricate one ready when the window of opportunity has closed. An operational artist has to time the planning, so that the quality of the plan and its completion converge optimally. Time cannot be wasted in drafting perfected plans. Thus,

> Don't Delay: The best is the enemy of the good. By this I mean that a good plan violently executed now is better than a perfect plan next week. War is a very simple thing, and the determining characteristics are self-confidence, speed, and audacity. None of these things can ever be perfect, but they can be good. (Patton 1947, p. 354)

Operational artists must settle for the good and cast aside hopes of attaining the best. Just *enough* time has to be used to plan properly, but no more. The operational artist cannot be a perfectionist since this would waste time.

The Prussian army was the first one in Europe to plan its wars thoroughly beforehand. von Moltke was the predecessor of modern planning processes and validated his plans in three wars over six years (Bucholz 2001, p. 10). It was a huge leap onward, but his times were different from ours. It is insufficient today to plan in detail if the plan is unadaptive to alterations caused by fluid situations. Independent, analytic, and occasionally unorthodox action is required from the commander. An operational artist cannot rely on standard operational procedures, but has to invent. At least in theory most military minds agree. "Theory is capable of benefitting only those who have raised themselves above the fray and have become completely dispassionate [...] A narrow doctrine would probably confuse us more than guide us" (Svechin 1992, p. 62). To win time the commander must extrapolate from the doctrine. There is a temporal discontinuity between the doctrine and its flexible application in the battle. MacArthur (1964, p. 79) wrote how the type of officer before the World Wars was unable to wage mechanized wars. They had "developed to handle a more or less recalcitrant element along definite and simple lines, and a fixed psychology resulted." Fuller argued that the rapidity of action in mechanized warfare could increase five- or tenfold and

> therefore each hour we have to-day to plan in, modify a plan in and issue orders and instructions in, will be reduced to from twelve to six minutes. A fixed idea is out of the question, the idea of the plan must be flexible—that is, it must embrace a number of alternative actions. (Fuller 1943, p. 48)

While the numbers were faulty before WW II and even further off their mark today, the speed of maneuver sets the pace that decision-making must reach. The quicker the troops the less time there is to control and direct their maneuver. Since there is less time to plan and command, time has to be relocated for these purposes from other spheres of action. Planning has to commence even earlier and it must create several different potential courses of action to be taken should the need arise.

We could say that there are fundamentally two different approaches to planning. One emphasized how the plan should be very detailed but temporally extend only to the initial contact. This type of thought is exemplified by von Moltke, for whom the first contact between enemies made the rest of the plan useless. After that Simpkin (1985, p. 14) described the continuation of the battle as "matter of responsiveness and opportunism" and "the right of the commander on the spot to react to situation as he found it." This is more or less the type of thought the Germans revitalized for WW II after adhering to the other type of thinking that Schlieffen propagated. He sought to "make an operational plan carrying out from mobilization right through to the strategic decision" (Simpkin 1985, p. 15). Neither one of these ideas is applicable per se and Simpkin exaggerated the rigidity of thought in them. Originally von Moltke had argued that one must "avoid planning beyond the situations one can foresee. These change very rapidly in war. Seldom will orders that anticipate far in advance and in detail succeed completely to execution" (von Moltke 1993, p. 184). de Jomini made the same point even earlier. For him meticulous planning was a prerequisite of success in battle, but he conceded that things change and so must plans. Everything must be done to ensure that the commencement of a battle occurs according to plan. "Up to this point everything relates to a first plan of operations; but no plan can provide with certainty for that which is uncertain always—the character and the issue of the first conflict" (de Jomini 2007, p. 244).

de Jomini and von Moltke were too wise not to ensure that there would be a mental outline of a plan how war should progress to the strategic decision. The Schlieffen Plan was not detailed all the way to the end but a sketch. The plan aimed at a rapid strategic decision in France. Just because the strategic time frame for victory over France was supposed to be short, the plan had to extend to the end of the operation. Still, as it turned out, the Schlieffen Plan was insufficient to win strategic victory by quickly collapsing France while it came very close (Corum 1992, p. 2). The plan was inflexible and didn't provide enough alternatives and contingency plans to respond to the emerging situations. It lacked plasticity because it was based on well-established strategic principles that had long ago hardened into rigid doctrine in Schlieffen's mind (Ritter 1958, p. 77).

Finkel (2011, p. 2) has argued that flexibility is one of the keys of coping with surprises in warfare. Flexibility is a combination of doctrinal, cognitive, command, organizational, and technological elements. We have already discussed most of these in passing, but will now focus on cognitive and command elements. Flexibility of mind and plasticity of plan enable the commander to save time when the plan and the reality drift apart. Franks (2004, p. 379) wrote that no matter how detailed a plan is, the reason why it is unlikely to survive the first clash is that "in any war plan the enemy gets a vote." When the enemy derails the plan, the commander causes unpredictable damage if he refuses to admit and accept the growing chasm between wishful thinking of the plan and the unfolding situation. As de Saxe (1944, p. 97) wrote, "sometimes these things which change the situation in question so greatly are overlooked until they are forced on your attention. Then it is too late, and you see yourself reduced to being ridiculous." As horrifying as the prospect of looking ridiculous is to a high-ranking officer, such personal considerations are inconsequential. If the commander is set in his preconceived ideas and a detailed plan that doesn't allow for improvisation, the entire operation is endangered.

> A mind that adheres rigidly and unalterably to original plans will never succeed in war, for success goes only to the flexible mind which can conform at the proper moment to a changing situation. This is what Napoleon meant when he said he never had a plan of operations. (von Freytag-Loringhoven 1991, p. 273)

"*Je n'ai jamais eu un plan d'operation*" should not be an operational guideline. Flexibility of mind is required for plasticity of plan to become reality and both can be attained only when the minds of the commander and his staff are fine-tuned to detect changes, seeking for the moment when the plan and the actual events start to diverge. This is a ***kairos***-moment for quick decisions when time won increases in importance. Some have taken Napoleon's words verbatim, but this is a false interpretation. Even if he confessed to having no 'plan', he constantly thought about all eventualities and never lost sight of his ultimate objectives. von Bernhardi (1914, p. 167) wrote how "he alone who has well thought out the art can practice it." Napoleon sought to imagine three or four months beforehand what could be the worst thing to happen (von Freytag-Loringhoven 1991, p. 287; Fuller 1961, p. 47; Goltz 1906, p. 103). He studied and thoroughly prepared the road that would lead to success. Napoleon claimed that "it is my habit to take so many precautions, that nothing is left to chance" (Cited in Fuller 1961, p. 47). He was not concerned only with what he should do but more on how the enemy's designs could affect his operations. Napoleon acknowledged that because of the changing circumstances the means of securing the objective in the mind of the commander "can never be sketched out with certainty long beforehand" (Goltz 1906, p. 103).

Nevertheless, Napoleon didn't conduct a war whimsically. In Napoleon's campaigns planning, marching, and the battle itself were parts of the whole (Smith 2008, p. 36). Furthermore, his every decision was based on intellectual evaluation *of any given moment.* The grand scheme was thought out but the plan of operations developed step by step, never seeking looking beyond the ongoing battle and quickly adapting to each situation. "Certainly the commander in chief *(Feldherr)* will keep his great objective *(Zweck)* continuously in mind, undisturbed by the vicissitudes of events. But the path on which he hopes to reach it can never be firmly established in advance" (von Moltke 1993, p. 45–46). One should enter every battle with a vision of how to continue and a 'blueprint' for the next phase of the operation. Napoleon attempted to play all the possible scenarios of development in his mind and chose an active path to meeting the emerging challenges in the course of his operations. He was a realist not prone to the failure of commanders who suffer from

> the tendency to be over-optimistic. The consequence usually is that the enemy is able to achieve a surprise, if not against the troops themselves, then against the commanders. A commander who fails to accept warnings, facilitates the winning of a great victory—for the enemy. An erroneous appreciation of the situation is an essential factor of defeat. In recent wars, many successful surprises were made possible by the incredulity of commanding officers. (Ehrfurth 1991, p. 419)

Napoleon didn't mentally chart the easiest way to success but the most difficult alternative routes the enemy might force him to take. One is not able to image every possible future development. The key issue is to focus on the probable ones since, "the most probable eventualities can be foreseen, however, because they depend on well-known and permanent conditions" (von Moltke 1993, p. 100). Napoleon was not easily surprised but due to his realistic outlook of future eventualities able to repeatedly cause surprises. Sikorski found the way to combine the fact that Napoleon on one hand lacked a plan of operations and on the other planned his operations far ahead. He described how a commander-in-chief should have

> simply a general idea of the operations (and not a plan) in connection with which will be formulated the most flexible possible plan of concentration. He will study the principal hypotheses which could present themselves in case of a war and the manoeuvres which he could arrange in each of them, without, however, fixing his intentions a priori. (Sikorski 1943, p. 168)

To put it in other words, the plan should be so intangible and unformed that it is rather a mental sketch of possible circumstances that might occur and how to maneuver in each of them. A flexible plan consists of outlines of several plans each one of which can be filled with detail should the situation call for it. "In the reality of war, things always turn out differently from what was originally expected and nothing is more natural" (Goltz 1906, p. 103). One way to ensure the survival of the operational plan beyond the first clash is to include different options to choose from in response to changed situations. This can be accomplished by creating what Pierre de Bourcet called a "plan with branches." According to him every plan of operations should have several branches so well thought out that one or another would be viable to employ in every situation. Flexibility of the plan has to allow the choice of the most suitable branch (Vego 2009, p. IX–121). Fuller argued on behalf of the same flexibility when he called for orders

> based on a profound appreciation of possibilities and probabilities, which, as I have explained, will generally lead to a series of alternatives. Therefore an order should not be suited to one operation but to several possible phases of this operation. It should possess a central idea and several radii working out towards the final circumference. (Fuller 1943, p. 158–159)

Some situational changes and surprises can be discarded as inconsequential while some others have to be responded to. The skill to discern between the two requires a clear mind from the commander. The commander must adapt to rapid changes and amend his plan accordingly, since

> adaptability is the law which governs survival in war as in life—war being but a concentrated form of the human struggle against environment. While the commander may initially decide to seek alternative objectives, if the enemy concentrates to cover this he will be wise to strike the other, more exposed. A plan must have branches like a tree if it is to bear fruit. (Liddell Hart 1932, p. 93)

Both the mind and the plan of the commander must be on one hand flexible and adaptable and on the other hand resolute and steadfast. Svechin warned about walking a tight rope between making a plan that doesn't leave too much to chance but doesn't "get bogged down in details and or delve too deeply into the various scenarios that may be encountered in carrying out the plan" (Svechin 1992, p. 279). Branches are responses to these scenarios. Branches can be viewed as "options built into the basic operation plan for a campaign or major operation. They are, in fact, contingencies within a given phase" and their ultimate purpose is to "allow the operational commander to anticipate future enemy actions that might lead him to drastically modify his basic plan" (Vego 2009, p. IX–122).

The branches are pre-planned means of modifying the plan when something unexpected occurs. This type of mental preparedness is an operational insurance policy. Even if battles are victorious, this doesn't mean that the operation would necessarily continue according to the plan. Every changed circumstance is reflected in future possibilities. Within an operation

> each battle changes the situation as completely as a twist does the coloured glass of a kaleidoscope. Thus the general is compelled every day, and often within a period of few hours, to modify his plans to suit fresh situations. (Goltz 1906, p. 105)

The plan of operations must be a sketch to which details can be added when effects of enemy actions and their influence upon the situation are assessed. Thus, the plan is never final but continuously evolving. At all times it should be the basis for action. The commander cannot

> act at random. Each operation has a raison d'etre, that is, an object; that object, once determined, fixes the nature and the value of the means to be resorted to as well as the use which ought to be made of the forces. (Foch 1920, p. 13–14)

If an operational commander doesn't focus his intellect beyond ongoing combat, he is carried helplessly in the torrents of the flow of time. An objective or an end-state for the operation needs to be set and bound to a time line to be the one and only firm reference point in warfare when all other things change constantly. However, for Foch too many additional things in plans were fixed and immutable. MacArthur (1964, p. 57) wrote that while Foch was a great general he was also "too inflexible once he had outlined a plan, and consequently misses opportunities." The ultimate goal expressed as the commander's intent is the only permanent thing that can provide a rationale for planning and executing operations.

The end-state cannot be too abstract and too far in the future or it becomes unattainable. It would be like the horizon; with every step one takes, it moves in turn that one step further. If the sights are set too close, the anticipations are so thin and short-range that change always catches one surprised and confused. *Coup d'oeil* must conjoin with powers of imagination and intellect and then the commander as "the adaptive individual appears to be able to project himself forward just the 'right' distance in time, to examine and evaluate alternative courses of action open to him before the need for final decision, and to make tentative decisions beforehand" (Toffler 1990b, p. 420). Plasticity of the plan means being flexible in relation to timings and temporality in general. Every operational plan has to be

> flexible enough to provide sufficient reserve time if something goes wrong or the action takes more time than anticipated. The more objectives or tasks assigned, the more time is required to accomplish them. Therefore, it is critical to focus on the most essential objectives, or tasks that will collectively ensure the accomplishment of the ultimate operational or strategic objective. (Vego 2009, p. III–26)

By endowing the plan with temporal flexibility, that is, the ability to survive altered timings, the life span of the plan may be extended. However, the more objectives, the more time is required to reach them and during each phase every moment of delay accumulates and is multiplied before the next objective is reached. Therefore, objectives need to be prioritized so that if the operation proceeds slower, the less important ones can be discarded. As Tukhachevsky argued, "battle must not be seen as some kind of smooth-running conveyor belt on which the various technical combat resources are merged. Battle is a complex and fickle thing. So command and control must be ready to deal with abrupt changes in the situation, and sometimes to reshape an earlier plan radically" (Cited in Simpkin 1987, p. 149). The same applies to operations that require even more flexibility. The plasticity of plan makes it possible for the operational artist to improvise because

> even a splendid plan cannot always be carried out successfully. A plan must be flexible and avoid the idea of adhering to certain schedules. In carrying out a plan we must be ready to take advantage of all the favorable opportunities presented to us and the enemy's mistakes at any moment. (Svechin 1992, p. 279)

Occasionally, when everything goes according to plan but the enemy does something unexpected, for the planned action to have impact, the plan has to be modified in mid-stride. This can happen even when progress is smoother than the plan presupposed. For Norman Schwarzkopf such a situation occurred in the Gulf War when the Iraqi units started to pull out of Kuwait City. Coalition operation proceeded smoothly, but this action of the Iraqis would allow them to suffer less damage. That was when he knew he

> had to act. Timing is everything in battle, and unless we adjusted the plan, we stood to lose the momentum of the initial gains. I'd fought this campaign a thousand times in my mind, visualizing all the ways it might unfold, and from the fragmentary reports coming into the war room I could discern that the Iraqis were reeling. If we moved fast, we could force them to fight at a huge disadvantage; if we stayed with the original timetable, they might escape relatively intact. (Schwarzkopf 1993, p. 525)

Flexibility is a requirement of the plan and the planners mind alike. The latter must be prepared to reformulate the former should the need arise. To save time, one should only alter the plan and not create a new one. The more elaborate the plan, the longer the time required to formulate it and consequently the more elevated the threshold of discarding that particular plan and writing a new one. The egress from this loop can be found in the initial draft of the plan; it needs to be relatively general and simple so that it can be amended with more detail as the situation becomes clear. Patton (1947, p. 248) wished to further underline how the flexibility of command manifests itself; "successful generals make plans to fit circumstances, but do not try to create circumstances to fit plans." Attempting to create favorable circumstances is the operational artist's job but in warfare omnipresent changes in the situation must be accepted and adapted to. Flexibility of plan can be attained through simplification but flexibility of mind requires complex thinking. Svechin called for simplified plans by saying that,

> the form of an operation should be as simple as possible. Piling up layer upon layer of battles and maneuvers in an operation is unacceptable not only because it makes command more difficult but because any excess maneuver or any battle which is not unavoidably necessary for achieving the goal on an operation holds the grave danger of distracting us from the goal. (…) Nothing should be superfluous in an operation because it should be the embodiment of purposefulness. In terms of its precision, clarity and symmetry the form of an operation should remind us of the straight lines of a Grecian temple rather than the swirls and whirls of Rococco. (Svechin 1992, p. 280–281)

However the demands of simplification and comprehensiveness of the plan are balanced one maxim of the mechanized age still remains true. When it comes to the operational commander, "his plan must never crystallize, for the energy of the battle front is always fluid" (Fuller 1923,

p. 35). Since both armies constantly alter their course of action in reaction to enemy activity, the plan must present only a general idea of desired development. Fuller used the expression "plasticity" to refer to the necessity of a plan being adaptable. According to him,

> as every policy must be plastic enough to admit of fluctuations in national conditions, so must each plan be plastic enough to receive the impressions of war, that is power to change its shape without changing or cracking its substance. This plasticity is determined psychologically by the condition of mentality in the two opposing forces. (Fuller 1923, p. 44)

4.7 Fluidity in Operational Art

> The Iraqis had fielded more men, more tanks, and more artillery pieces than the coalition. What defeated them was the combination of superior technology, realistic training, and the fluid, flexible tactics of the coalition forces. (Franks 2004, p. 165)

The vision of fluidity in the use of military force extends over the timespan from ancient China to contemporary Western forces Franks commanded in the Gulf. There still remains the need to aim even further than flexibility or plasticity and try to increase fluidity in operational art. Sun Tzu with his ideas of armies operating like water has been rephrased over and over again by proponents of maneuver theory. The two most common metaphors used are the fluidity of operations in what Liddell Hart termed "indirect approach" of finding openings in enemy defenses and seeping through them like water into the depth or the idea that a superior force exerts pressure on the enemy defense and after a breakthrough the attacking forces will flow irresistibly like through a busted dam into the depth of the enemy (Danchev 1998, p. 155–163). An example of the latter type was used by Guderian who described the early stages of WW I by saying that "the finest army in the world had flooded like a wall of water across the Meuse and deep into the enemy country to the south. Two months later, when the leaves were falling in the autumn of 1914, the grey tide was ebbing" (Guderian 1992, p. 33). The former type is exemplified by Liddell Hart's argument that such a breakthrough can, with sufficient pace, provide "a decisively deep penetration so long as it can be kept up by a torrent-like process of advance, either swerving round resistance or piercing it as a weakened spot—in which case the tank-torrent contracts in pouring through a narrow breach, and then expands again to its original breadth" (Liddell Hart 1950, p. 272). He attempted to coin a new term to describe the possibilities provided by mechanization. Even if his idea is connected to dispersed armored forces concentrating for attack, it could just as well describe more recent doctrine of AirLand Battle. An important characteristic of a mechanized attack was

> fluid, or distributed, concentration. To strike, by fire alone, at the greatest number of points in the shortest time over the widest area. And without ever making contact in the present tactical sense. Never giving the enemy a target, yet enticing him to waste his ammunition and keeping his nerves at an exhaustingly high tension. (Liddell Hart 1932, p. 212)

The idea of fluidity meant for Liddell Hart the possibility to concentrate force, but the way he described the idea of striking numerous targets from a distance could be from the texts of the NCW theorists. The essential idea is the same, but prior to WW II, there was no other choice than tanks and air force to carry out such an operation. Technology has since enabled long-range combat without troops getting into contact. In mechanized warfare tanks were the primary means to create this fluidity of maneuver. A mechanized force could also use its armor

> simply for a close approach, not for an attack; to move up to a "fluid" position, whence, in comparative security, it can smother the enemy or cut his arteries of supply by a demoralizing fire. (Liddell Hart 1932, p. 211)

It is not much of a leap of imagination to give credit for these thoughts to Sun Tzu, whom Liddell Hart read and who compared armies to water since "as water shapes its flow in accordance with the ground, so an army manages its victory in accordance with the situation of the enemy. And as water has no constant form, there are in war no constant conditions" (Sun Tzu 1971, p. 101). It was typical to Liddell Hart not to refer to his sources of inspiration or pay tribute to the original ideas (Yuen 2014, p. 128–138). He used fluid as a metaphor for his original contribution of "Expanding Torrent" tactics. Liddell Hart argued that it is a natural way of attacking because watching

> a torrent bearing down on each successive bank or earthen dam in its path, we see that it first beats against the obstacle, feeling and testing it at all

> points. Eventually it finds a small crack at some point. Through this crack pour the first driblets of water and rush straight on. The pent-up water on each side is drawn towards the breach. It swirls through and around the flanks of the breach, wearing away the earth on each side and so widening the gap. Directly it has passed through it expands to widen once more the onrush of the torrent. (…) Thus nature's forces carry out the ideal attack, automatically maintaining the speed, the breadth, and the continuity of the attack. Moreover, the torrent achieves economy of force by progressively exploiting the soft spots of the defence. (Cited in Danchev 1998, p. 103)

Characteristically Liddell Hart first thought of the method and then created a parable in the shape of a small narrative to make it alive in the minds of his audience (Danchev 1998, p. 176). He saw that a fluid concentration of force might provide an escape from entrenched fronts and battles of attrition. He claimed that "fluidity of force may succeed where concentration of force merely entails a helpless rigidity. The sea is stronger than a steam-roller, and should replace it as our military ideal" (Liddell Hart 1937, p. 91). Sun Tzu had long before argued that the army should resemble water in its movements because

> the nature of water is that it avoids heights and hastens to the lowlands. When a dam is broken, the water cascades with irresistible force. Now the shape of an army resembles water. Take advantage of the enemy's unpreparedness; attack him when he does not expect it; avoid his strength and strike his emptiness, and like water, none can oppose you. (Sun Tzu 1971, p. 89)

Even if the army flows in fluid motion in its offensives, it doesn't lessen its power, because the force can be accumulated and unleashed in a simultaneous and preferably surprising wave of attack to overwhelm the defender. Fluidity entails for both Sun Tzu and Liddell Hart that the dispersed formations coupled with mobility make an elusive target for the enemy. You cannot punch a hole in water or hurt it, but water can fill every hole, nook, and cranny in the defensive formation when the wave sweeps over them. The idea of fluidity of operations meant that an attack

> should follow the pattern of flowing water. As water proceeds downhill, it naturally avoids strong surfaces. Instead, it flows about seeking weak points and gaps through which the water begins to trickle. When such gaps are found, the whole body of water rushes toward it, speeds through it, and

> then expands on the other side. So also an attack should avoid enemy strengths (the surfaces) and exploit the weak spots (the gaps). Once through the gaps, the attacking force expands to destroy critical enemy units and installations. (Leonhard 1991, p. 50)

Liddell Hart's call for fluidity of force seemed original in the Occident, but theorists in the Orient had read their Sun Tzu as well and perhaps even more thoroughly. As an example we can use Mao, for whom everything was fluid in warfare.

> Fluidity of battle lines leads to fluidity in the size of our base areas. Our base areas are constantly expanding and contracting, and often as one base area falls another rises. This fluidity of territory is entirely a result of the fluidity of the war. (Mao 1963, p. 136)

Fluid operational art would enable asymmetric forces to disperse in front of an enemy attack. They don't defend territory, but allow enemy to take over and recede like water in front of it only to build up as if behind a dam somewhere else to cascade over the enemy at the time and place of their choosing. Liddell Hart and Fuller promoted mainly flexibility of command and thought in the West, but the Israeli army built a whole doctrine of "continuous flow" into their operational art and for most of the Cold War era was the main proponent of maneuver warfare (Bond 1998, p. 10–11; Kesseli 2001). Flexibility implies the presence of some superstructure providing shape and support, but that it can bend and be reshaped in response to altered situations. No matter how flexible an object is, it possesses more internal structure than fluid. Fluidity enables freer movement and even metamorphosis, but entails stability and the option of solidifying, if required. Because of

> the fluidity of war, some people categorically deny that war plans or policies can be relatively stable, describing such plans or policies as "mechanical". This view is wrong (…), because the circumstances of war are only relatively certain and the flow (movement or change) of war is rapid, war plans or policies can be only relatively stable and have to be changed or revised in good time in accordance with changing circumstances and the flow of the war; otherwise we would become mechanists. But one must not deny the need for war plans or policies that are relatively stable over given periods; to negate this is to negate everything, including the war itself as well as the negator himself. (Mao 1963, p. 243)

While some Occidental writers occasionally allowed themselves to be carried away by their ideas, Mao kept a level head. He argued that everything is fluid, but simultaneously proposed that things have to be stabilized at least for a given duration to avoid chaos. There must be a means to control the impalpable nature of fluid. The operational plan can provide stability for the fluidity of war if we treat it as a dam that keeps the fluid from flowing freely for a pre-determined period of time. But how long is this period of relative stability the plan provides?

> The period of validity of a plan for a campaign is shorter than that of a strategic plan, and for a tactical plan it is shorter still, but each is stable over a given period. (…) No one denies that even a plan valid for a given period is fluid; otherwise, one plan would never be abandoned in favour of another. But it is fluid within limits, fluid within the bounds of the various war operations undertaken for carrying it out, but not fluid as to its essence; in other words, it is quantitatively but not qualitatively fluid. Within such a given period of time, this essence is definitely not fluid, which is what we mean by relative stability within a given period. In the great river of the war as a whole, in which fluidity is absolute, there are various stretches, each of which is relatively stable. (Mao 1963, p. 244)

Periodically the fluid nature is restricted to attain temporary stability. This doesn't occur by solidifying the fluid but by letting the plan act as a superstructure that contains fluidity of operations by restricting their flow. Flowing fluid is brought under control by giving it a direction like riverbanks do. Mao wrote of relatively stable stretches in the river of war, but in any river there may be strange undercurrents. Mao advocated flexibility of plans in the manner of Sun Tzu's original idea. He had used water as a metaphor to emphasize the idea that no general should follow a predetermined plan (Yuen 2014, p. 80–81). That is, plans were not supposed to be rigid and unchanging even if they provided the temporal containment of the fluidity of warfare. We might be even better off by viewing the plan not as the banks of a river but something with even more plasticity, like a water balloon filled with the fluid essence of war. Because everything in warfare is fluid and stability only fleeting and transient, the structure provided by plans must change shape as well. Therefore the plan

> must change with the movement (flow or change) of the war and vary in scope according to the scale of the war. Tactical plans, such as plans for attack or defence by small formations or units, often have to be changed

> several times a day. A plan of campaign, that is, of action by large formations, can generally stand till the conclusion of the campaign, in the course of which, however, it is often changed partially or even wholly. (Mao 1963, p. 243)

Change is the only constant factor in the course of war and thus operations, battles, and plans must abide to it. It would be insufficient to emulate the plans of past great captains, since when their methods have been thoroughly analyzed, what once was fluid and changing is solidified into a doctrinal form. The originality is lost in repetition and in being formalized into an accepted practice. The truly unorthodox cannot be put to words and turned into dogma.

> What has form and outline will be seen and praised by the world; what has chapter and verse will be transmitted and studied by the ages.
> These are all examples of forms overcoming one another. The one who is skilled at form does not use them as a model. What ennobles the Way is its formlessness. Having no form, it thus
> cannot be controlled or coerced;
> it cannot be measured or ruled;
> it cannot be tricked or deceived;
> it cannot be schemed against or planned for.
> People will make plans for one whose wisdom is apparent;
> they will attack one whose form is apparent;
> they will ambush one whose numbers are apparent;
> they will defend against one whose weapons are apparent. (Liu An 2012, p. 110)

Fluidity is about being original and not adhering rigidly to any form of tactics or operational art and because of that surprising the enemy by being unexpected and even unexpectable. The requirement of fluidity extends from the fighters to the people who support them. One of Mao's most famous quotes tells us that "the people are to the army what water is to fish, as the saying goes" (Giap 1962, p. 56). The insurgent army requires support of the people to survive. The fighter hides among the people from his enemies, since finding a single fish in the sea is almost impossible. The soldier in a people's war can be compared to both a fish swimming among the population and a man swimming in the sea. This applies to other forms of asymmetric warfare as well. To handle the fluid and ever-changing nature of the 'sea' of war, the soldier must be able to clearly determine how to get across it and do every necessary thing with deliberation. Directing and fighting a war are like swimming. To quote Mao, again, he argued,

> Swimming in the ocean of war, he not only must not flounder but must make sure of reaching the opposite shore with measured strokes. The laws for directing war constitute the art of swimming in the ocean of war. (Mao 1963, p. 87)

4.8 Timely Information and Time-Lag in Decision-Making

> A commander's correct dispositions stem from his correct decisions, his correct decisions stem from his correct judgements, and his correct judgements stem from a thorough and necessary reconnaissance and pondering on and piecing together the data of various kinds gathered through reconnaissance. He applies all possible and necessary methods of reconnaissance, and ponders on the information gathered about the enemy's situation, discarding the dross and selecting the essential, eliminating the false and retaining the true, proceeding from one thing to another and from the outside to the inside; then, he takes the conditions on his own side into account, and makes a comparative study of both sides and their interrelations, thereby forming his judgements, making up his mind and working out his plans. Such is the complete process of knowing a situation which a military man goes through before he formulates a strategic plan, a campaign plan, or a battle plan. (Mao 1963, p. 84)

This planning and decision-making process that goes on in the head of the commander that Mao described is comprehensive and if the commander is intelligent, likely to produce a feasible solution to the dilemma facing him. But this process takes time even in those rare cases when the commander is a genius. Time is a scarce resource that one has to manage meticulously. Operational artists must be acquainted with

> how to utilize time to the best advantage, and this demands a perfectly organized instrument in which friction, which is the enemy of military time, is reduced to its lowest possible level. To understand the time limitations of

> one's own side and of the enemy's is to work from the surest of foundations, and if our organization will enable us to move more rapidly than the enemy, then from the start we possess an immense advantage over him, for indirectly this organization will enable us to increase the time at our disposal. (Fuller 1926, p. 179–180)

Fog of war slows down commanders in their decision-making and friction of war the movement of their troops. Thus, friction truly is the "enemy of military time." Not only time is squandered but energy of movement is consumed in overcoming the friction. The relationship between fog and friction is important. Both are permanent elements of battlespace. Friction complicates maneuvers since it makes movement in warfare resemble movement in water or other resistant element (Howard 1983, p. 26). Friction is caused by effects of time, space, and human nature and for von Clausewitz it is the unavoidable force that makes actual war something less than the ideal of absolute war. Events take time to unfold and friction increases this duration (Bassford 1994, p. 25). To define the Third Wave meanings of fog and friction NCW theorists provided a suitable starting point. They argue that "the fog of battle is about the uncertainty associated with what is going on, while the friction of war is about the difficulty in translating the commander's intent into actions" (Alberts et al. 2000, p. 71). This definition, however, entails only half of the concept. This one-sidedness stems from the focus on the characteristics of the information age. They only refer to the mental, intellectual, and immaterial aspects of fog and friction in information-related issues.

It would be wiser to consider the mechanical and physical aspects as well, since fog and friction had their widest manifestation during the indust-reality. von Clausewitz excluded them since he favored moral factors of war over the physical ones but mechanization made the physical aspects of operational art to regain importance (Howard 1983, p. 26). Therefore, friction doesn't reside only in the mind of the commander or the mental activity in the chain of command before an order transforms an intention into action, but also in all physical activity related to initiating and upholding movement within the battlespace. Friction is unavoidable and an omnipresent element in war (Gray 2007, p. 74–75). It slows the execution of plans, but it is not necessarily fatal. One must temporize his actions, so that the effect of friction is included in the rhythm of operations.

Fog can be understood as lack of adequate information for decision-making and friction consists of all internally or externally caused glitches of immaterial or material origin in the war-machine that delay operations. The classic Clausewitzian interpretation of friction doesn't so much bog down operational mobility in the physical battlespace as cause time lag in mental conception of the battlespace. Today speed is limited by physical restrictions imposed by technology and actions of the enemy but also by the mental capacity of the soldiers. The cogwheels of the military machine occasionally grind to a halt because of mental friction since, "every soldier feels that there is something clogging the mechanism of manoeuvre, even though he may be puzzled as to what it is. Missed opportunities remain very common, and operations often get stuck for some reason other than the enemy's opposition" (Liddell Hart 1950, p. 290). Friction occurs on both physical and psychological levels and while in the former planning and timing the movements of the units properly reduce friction in maneuvers, the latter is far more challenging to address.

Leonhard (1998, p. 49) rejoiced that acquiescence to the fog could somehow end because "twenty-first century holds out a new and exciting possibility: that future leaders can harness information to the advantage." This optimistic thinking resulted from excessive faith in technology. Fog will not evaporate since "more advanced communication would not change matters greatly, as the evidence of recent wars has shown. As soon as movement begins, so does the fog of war" (Luttwak 1987, p. 106). Movement reduces the effect of better communications and information technology. The fog of war is dispersed in stasis but as soon as movement starts, the fog begins to condensate again and obscure visibility. "The bulk of the fog will remain just as dense, for increased mobility will cause situations to change rapidly" (Fuller 1943, p. 55). The greater the mobility of troops the shorter is the time intelligence information remains accurate. Yet, efficient maneuver is a prerequisite of operational art and one must set priorities. Moving fast is still often more important than perceiving clearly the prevailing conditions. We can compare effects of speed on detailed situational awareness to a view through the window of a train; the faster the speed, the more blurred the vision.

NCW doesn't deny the omnipresence of fog and friction as elements of warfare. It attempts through sharing a "better near real-time picture of what is happening" to increase battlespace awareness to reduce their effects (Alberts et al. 2000, p. 11). Fog affects commanders on all levels. When they cannot 'see' either with their own eyes or through intelligence data, battle reports, or other external sources what occurs within the battlespace, they have a choice either to be quick to decide and risk making erroneous decisions or wait for sufficient information and lose time. Increased awareness supports decision-making, but there is a risk of senior commanders on operational and strategic levels starting to micromanage battles as the fog is lifted. This would only increase friction (Kagan 2006, p. 209).

Friction affects every activity within the battlespace from the physical maneuver to the mental such as decision-making and NCW claims to reduce friction through increased responsiveness inbuilt to the network structure itself. This, in theory, allows for self-synchronization and autonomous responses that would save time, since decision-making is partially outsourced to the network. In other words, while friction remains a part of warfare, the network should reduce human friction through autonomous and knowledgeable networked entities with automated responses to specific and pre-determined stimuli. The capabilities of the network are used to support decision-making, and as Alberts et al. (2000, p. 12) claim, "there is definitely a place for automated tools and decision aids on the battlespaces of the future."

There is today a pressing need to properly define the place and the tasks of these automated tools. "New technologies offer both promise and a potential problem. Higher commanders and political military leaderships should not be tempted into believing that technology can or ever will satisfactorily resolve all the problems associated with operational command and control" (Vego 2009, VIII–21). Echevarria (2013, p. vii) seems to agree, since he wrote about the U.S. attempting to use high technology to gain knowledge and situational awareness but ultimately finding it still impossible to thin the fog enough to eradicate the role of chance. Fog of war may thin on occasion and the Third Wave offers tools to attempt to pierce it, but even with the latest information technology exists the "danger of knowing more and more about oneself and proportionally less and less about the enemy" (Smith 2008, p. 411–412). It always has been difficult to get precise and timely information on the disposition of the enemy. Were it not so, warfare would be easy even with inferior forces (Ludendorff 1919, p. 101). It is not important to get information as soon as possible, but to get in time for analysis so that the results can be incorporated into the operational plan. One needs information at intervals synchronized with one's own activities.

"In war nothing is ever as bad, or as good, as it is reported to higher headquarters" (Patton 1947, p. 354). Information is like Guinness draft beer; it doesn't travel well. To get the most of it, one must access it directly at its source. Rommel belonged to the school of generals who wanted first-hand information. If there was any means of obtaining information on the spot he chose to go there and not rely on reports. As a drawback he often was out of reach of his staff when decisions had to be made. But Rommel prioritized things differently. As he wrote,

> for the commander to have a good understanding on the battlefield of his own and the enemy's dispositions is of utmost importance. It is often more important to have an accurate overview of the actual battlefield than to be intellectually more qualified, or to have more experience. This is especially true of a situation where developments cannot be foreseen. A man must observe and learn for himself, since reports from second-hand sources cannot be relied upon as a base for important military decisions. (Rommel 2003, p. 68–69)

The proximity of command to the fighting troops was characteristic to German panzer-officers but this stemmed from a long tradition. No matter how vast the battlefield, there always is one single spot where "the plot laid by the strategic and tactical conditions will thicken to a crisis. That is the point where the director of battle must be also found in the future. Here his personal intervention may be of decisive importance" (von Bernhardi 1914, p. 118). Guderian (1956, p. 231) was always close to the action and his soldiers to ensure that his ability to make sound decisions. According to him, no panzer-general ever stayed too far from the front. By placing the commanders in the front the Germans reduced decision-making time by erasing the time lag caused by relaying information. This enabled the tactical and occasionally operational commanders to survey the immediate conditions of combat and to take timely action. The commander needs to do his utmost to diminish the amount of decision-making based on unreliable information.

Schlieffen (1936, p. 91) agreed with his claim that "in war there is, however, nothing more dangerous than 'reliable information.' It is overtaken by stark reality. What was applicable yesterday may be wrong tomorrow." Not only information needs to be fresh, but actions must be undertaken without wasting a moment in processing it. Information is always insufficient and Ehrfurth (1991, p. 437) argued that "three quarters of the facts which one should know in order to make the right decision remain shrouded in uncertainty. He who waits too long for better information risks the loss of a good opportunity." Opportunities are fleeting and to make the best of them, the commander must use his *coup d'oeil* to produce a comprehensive mental picture of the situation. Only the outcome will tell if the foresight was shortsighted. Time waits for no man and commanders often have to make their decisions in the spur of a moment. According to Franks, to be a general in a command position "meant gathering as much information—always in short supply—as possible, then making decisions. And living with those decisions. It meant using judgment" (Franks 2004, p. 148).

The most damaging time lag in command occurs between gaining sufficient information to make a decision and making it. Shortening it would require training the commanders to make decisions faster and since demand would exceed supply of cognitive capabilities, it may prove impossible. The first Gulf War can be used to illustrate this time lag. A huge deal of information through various channels from satellites to operators on the ground was continuously streaming in, but analysis was slow and hand-delivery of information meant that in worst cases it took up to two weeks to get the results of the analysis to the units requiring them. By the beginning of the air campaign the delay had been cut to thirteen hours (Toffler and Toffler 1995, p. 80). Still the process of air sortie planning and targeting was considered too rigid and after the war the system was replaced. In the Gulf the land commanders got feedback from the air components slowly. They nominated possible targets; air commanders prioritized them and decided which would be hit. This process combined with 72-hour battle damage assessment cycle made it impossible for the land commanders to plan the next iteration of targets without considerable losses in time (Vego 2009, p. V–110).

In the land domain surprise was affected on Iraqi troops and commanders not by a particularly devious stratagem but the speed of movement with which the turning maneuver was executed. This was enabled not only by mobile forces, but computers, communications, and satellites provided the means to accelerate the velocity of warfare (Toffler and Toffler 1995, p. 79). The operational plan was no flash of Schwarzkopf's genius, but a well-executed enveloping maneuver with overwhelming superiority and great speed. The issue in battles and operations isn't

> necessarily absolute speed, but speed relative to the enemy's pace. And here there was no doubt about the speed superiority of the victors. (Ironically, the intelligence time lags would have been less troublesome if U.S. forces were not themselves moving so quickly.) (Toffler and Toffler 1995, p. 80)

Speed of maneuver influences other activities. In this case, the high mobility of the ground forces decreased the amount of timely information at their disposal by lengthening the time it took for analyzed information to reach them on the move. The speed of information flows would theoretically enable instantaneous responses, but as of today, humans make these responses and coordinate activities within the battlespace. All tempos of activities, be they of planning, supplying, moving, analyzing, producing orders, executing them, or even operations by various types of units, differ from each other. The most important thing is to synchronize speeds so that effects favorably conjoin in space and time. The activities of different types of units partaking in a joint operation may commence according to a time scale of their own but the results need to converge.

> The golden rule is so to arrange the sequence of initiatives that their effects converge both in time and space. Only on this basis can an action carried out in the various zones have the necessary coherence. This is in fact a wider application of the principle of economy of force. (Beaufre 1967, p. 81)

One of the great paradoxes of the Third Wave is that too much information adds to confusion. Just as during indust-reality masses of men and materiel caused congestion, today the mass of information may result in immobility unless it is properly managed. One should perhaps start to consider applying a principle of 'economy of information.' As we discussed in the case of force, there must always be enough of it but excessive amounts cannot be meaningfully employed. The flood of information commanders encounter is overwhelming. Many proponents of information warfare tend to confuse the concepts of information and knowledge (Smith 2008, p. 327–328). Information is valuable as a source of power but only knowledge is power. Turning information into knowledge takes time. The two concepts are not synonymous.

> Knowledge is information after it has been subjected to analysis, generalization, and utilization by application and abstraction. Sometimes what is knowledge is not informative, and what is informative is not knowledge. The real danger is to focus on knowledge as a commodity instead of as a dynamic phenomenon to be improved. (Vego 2009, p. III–67)

One must find the time and the people to analyze information. Only processed information is useful and raw data has to be refined into knowledge that gives the commander an edge on his opponent who may or may not have access to similar information flows. "Both knowledge and ignorance have dominated warfare throughout history, but Information Age warfare has adjusted the balance toward knowledge" (Leonhard 1998, p. 251). Operational artists have always made decisions based on little information of dubious accuracy. Excessive information flow may end up wasting instead of saving time. Information keeps flowing in, actions in the battlespace keep unfolding, and the commander may end up trying to endlessly master the 'now.' We tend to strive for complete information, but that is an oxymoron, because to have

> complete information does not necessarily mean being more informed; it can well lead to loss of orientation. What is needed is not more information but more orientation. One is often confused today, not necessarily ignorant. (Vego 2009, p. III–65)

Commanders should not be confused but determined and assertive. But as they struggle to create the perfect picture of the current moment from information received instead of employing their *coup d'oeil*, they always lag behind. As more and more information is collected and processed, "the result will be too much information, integrated poorly or not at all. The more information, the less its acceptance. The veritable flood of information today increases one's uncertainty about one's own opinion" (Vego 2009, p. III–66). If commanders are subjected to the information flow without filters blocking out the 'white noise' not crucial to the task in hand, their capability to make decisions will diminish. Just because there is so much information available, effort must, according to Smith (2008, p. 327), focus on the items and issues absolutely necessary to the commander. He defines his information requirements. Without an efficient analysis process taking raw data as input and rapidly providing knowledge as output, the fog of war thickens around the commander. Having information but lacking knowledge leads to uncertainties and second-guessing decisions. In the words of Leonhard (1998, p. 252–253), the acme of skill during the Third Wave is "*to manage what we know and what we don't know, and to balance our knowledge with activity.*"

4.9 Boldness as Time-Winner at Uncertain Times

Armed forces wish to save time in decision-making and officers are trained to make lightning-fast decisions even with a certain disregard of the eventual outcomes. This type of energetic and proactive approach to countering problems with instant solutions is valuable to teach, but it can potentially become a harmful part of the commander's mind-set. There is no room in operational art for thoughtless decisions, but the worst thing commander can do is nothing, since

> failure to act is worse than an error of judgment in selecting a course of action. Commanders in time of peace should therefore make it their duty to encourage initiative in their subordinates, instead of checking it, as is the case too often. Subordinates should be taken to task only where their action was taken thoughtlessly—without a good reason. (von Freytag-Loringhoven 1991, p. 216–217)

Rapid decisions and energetic execution enable operational artists not to waste time. Every officer throughout his career should be encouraged to decide and to act rapidly, but the most important lesson is in the last words; the time for every decision is only after it has been thought out and based on sound reasoning. If the original idea of consuming just as little time as is necessary in decision-making deteriorates into a fetish of applauding instantaneous decisions based on traditional demand for them, the risk of catastrophic blunders increases. The commander is exposed not as a great captain but an even greater buffoon. No matter how pressing the demand is, commanders need to produce informed decisions instead of automated ones.

Automation is permissible only in the absence of other choices. The operational plan seeks to identify in advance decision-making points along the path and suggest courses of action. No matter how bad a decision made on a spur of the moment is, if making one is required, then "of all faults, one only is degrading, namely inaction" (Foch 1920, p. 45). Inaction gifts time to the enemy as he can execute his respective plans unhindered. To fight effectively in any sense of the word demands decisions in large quantities. As Franks (2004, p. 479) described leadership of major operation, "once we committed the force to war, the challenges and decision-points never stopped."

In the agrarian era Leo VI (2010, p. 43) advised to "take your time in making your plans, unless some necessity requires immediate action. But once you have decided on something, unless there is an obstacle, carry it out quickly. In like manner, as was said, select the time and place and make preparations appropriate to the action." The commander chooses when and where he commits his troops into battle. In other words, "it is incumbent on you, O general, to take advantage of times and places" (Leo VI, 2010, p. 303). To take advantage is to play a high-risk game. Often the burden of decision is on the one who seizes the initiative. This led Caesar to declare "*Alea jacta est*" when he crossed the Rubicon. The die was cast.

When one chooses to initiate an operation, regardless of force concentrations of both belligerents and number-crunching mathematics that 'prove' numerical superiority and predict a victory, there are incalculable risks involved. No preparation can eliminate them completely. War is not an art for the timid and that is why accountants rarely lead armies to great victories. Risk will always be a part of warfare and

> he who will win everything, must dare staking everything on one single card. By doing so, a military genius does not act like a gambler. He acts rather as a serenely self-confident and bold personality who is inspired by the "sacred fire" of the will to win and who aspires to the highest success. (Ehrfurth 1991, p. 511)

Whether we call this inspiration and inner conviction along the lines of Montgomery or discuss it with terminology such as audacity, boldness, gambling, luck, or genius, only the active combatant can achieve a victory in war. Hopefully the "sacred fire" is more than heartburn. To quote von Freytag-Loringhoven (1991, p. 276), "however tactics change, they will not reduce the need for skilled generals who can act boldly on sketchy information." While this is an admirable feat of audacity, it is far removed from gambling. As Franks (2004, p. xiii) put it, "in thirty-eight years as a soldier, I'd learned the difference between a risk and a gamble." In uncertain situations

> bold decisions are the best way to success. Strategic and tactical boldness must be distinguished from a military gamble. Bold is that operation which, while having the possibility of success, also leaves one with sufficient forces in hand to cope with circumstances which might arise in the event of failure. A gamble, on the other hand, is an operation which can bring either victory

> or the total annihilation of one's forces. There are situations in which a gamble is justified, such as when defeat is only a matter of time, so that gaining more time is pointless, and the only chance lies in a risky operation. (Rommel 2003, p. 174–177)

Boldness doesn't equal gambling, since while his friends and foes alike appreciated Rommel for the audacity of his operations, he took into account the possibility of failure and left enough reserves to cope with if success evaded him. Rommel claimed that in Africa, he "had never gambled; even in the most daring operation, I had always kept enough in hand to deal with any situation, and had never had to fear losing everything" (Rommel 1953, p. 400). He reserved a time for gamble as well. This was when a defeat is certain and one can only decide when it occurs. To win time by delaying is useless, if time will not alter the situation. Only then a gamble can be initiated against overwhelming odds. When facing the inevitable, one must take the million-to-one chance. Fuller generally condemned gambling, but acknowledged that,

> by taking risks which are worth taking that, more often than not, the greatest economies are effected and the highest interest secured. In war, audacity is nearly always right and gambling is nearly always wrong, and the worst form of gambling in war is gambling with small stakes; for by this process an army is eventually bled white. (Fuller 1923, p. 36)

There is a time for boldness and a time for caution in war. Operational art is about balancing these two depending on the situation. "While tactical decisions tend to require a certain boldness, a strategic decision such as this should only be taken after meticulous examination of all possible consequences, and should, as a matter of principle, satisfy the need for 100 per cent security" (Rommel 1953, p. 348). In strategy security has predominance. Often the stakes are so high that making a decision calls for extraordinary mental strength because of the weight of responsibility. Mostly boldness belongs to tactics where by seizing initiative one is able to 'make it or break it' and extends into operational art. Without a proper strategic level plan that combines battles into operations directed toward a singular end-state tactics make no difference in the war. At the same time "the best strategic plan is useless if it cannot be executed tactically" (Rommel 1953, p. 389). Strategists and operational artists take different risks and when they gamble, the stakes differ. It is a question of personality whether the commander wishes to gamble or play safe. Hitler is often seen as a reckless gambler, but von Manstein painted a different picture of him. Hitler made the mistake of undertaking too many different objectives at the same time, being active everywhere and not daring to gamble by being strong at one place at a time.

> The rule that one can never be too strong at the crucial spot, that one may even have to dispense with less vital fronts or accept the risk of radically weakening them in order to achieve a decisive aim, was something he never really grasped. As a result, in the offensives of 1942 and 1943 he could not bring himself to stake everything on success. (von Manstein 1982, p. 276)

Hitler gambled like an amateur attempting to play several tables with insufficient stakes. For all the rashness he has been accused of, Hitler was too timid. On the other hand, Wavell, not often lauded as risk-taker extraordinaire, argued that risks need to be taken. For him the distinguishing hallmark between a great commander and an ordinary general was that the former had a

> spirit of adventure, a touch of gambler in him. As Napoleon said: 'If the art of war consisted merely in not taking risks glory would be at the mercy of very mediocre talent.' Napoleon always asked if a general was 'lucky'. What he really meant was, 'Is he bold?' A bold general may be lucky, but no general can be lucky unless he is bold. (Wavell 1953, p. 19–20)

During WW II the British commanders were generally more reluctant to give anything to chance while Americans often were more audacious. As Alexander of Tunis wrote, "Patton was a thruster, prepared to take any risks [...] Patton should have lived during the Napoleonic wars—he would have been a splendid Marshal under Napoleon" (Alexander 1962, p. 44). The idea of luck favoring the bold was characteristic to operational art everywhere until the World Wars. In pre-war Germany von Bernhardi (1914, p. 141) argued that "there is one quality above all in man which is of the utmost importance in all warfare, and really benefits the attack exclusively—boldness. Fortune smiles upon the bold commanders before all others."

In war nothing is certain, but having an overwhelming superiority in forces likely leads to victory if one avoids unnecessary risks. For the underdog, boldness and audacity provide the only chance to weigh the scales in one's favor. "We must not condemn a risk in general, but we must study beforehand whether risk is appropriate in a given instance. Only if the risk is inappropriate can we talk about an adventurer" (Svechin 1992, p. 265). Risks need to be evaluated and embraced if the situation so dictates. A choice between bold operations or playing it safe has to be done before an attack is initiated. The question before every operation is, "how should we do it? Strike swiftly and win swiftly, or strike surely and advance surely! This was the problem of the direction of operations in the campaign" (Giap 1962, p. 169–170). The operational artist "must be prepared to take a chance when the situation favours boldness. He will lose part of the fruits of victory if he is never prepared to soar from the known to seize the unknown" (Montgomery 1961, p. 49–50). Again, his greatest challenge is in locating the most suitable moment to test his wings. Whether he soars high or plummets like Icarus depends upon his choice of the right time to act.

Exceptional situations like having absolute superiority can justify taking the safe route. As Rommel (1953, p. 521) described the methods of Montgomery in Africa, "his principle was to fight no battle unless he knew for certain that he would win it. Of course that is a method which will only work given material superiority; but that he had. He was cautious—to my mind, excessively so—but then he could afford to be." Montgomery made full use of his material superiority and Fuller (1948, p. 235) argued that he "was pre-eminently a general of materiel." The scales were so much weighed in Montgomery's favor that he could proceed slowly and deliberately, step by step eroding the strength of Rommel's troops.

> The only time a commander can be sure of the outcome of a battle is when he has forces so superior to those of his enemy that a victory is self-evident. Then it is not so much how but when. But even in a situation like that, I believe it is better to pursue operations boldly. (Rommel 2003, p. 174–177)

Boldness implies accepting certain risks and if the risks actualize there is a chance of losing the battle. Montgomery's foe, Rommel, gave him credit for this by writing that while "command of a force in mobile battle was not his strong point [...] It would be difficult to accuse Montgomery of ever having made a serious strategic mistake" (Rommel 1953, p. 521). Montgomery had a great track record and took no risks because he didn't need to. He was deliberate and slow but he never lost a battle because he waged war in a thoroughly logistical manner, determined to give his forces every possible material advantage (Gray 2007, p. 116). With vast superiority there is no need for audacity, no need to win time. Similarly the Soviet offensives of late WW II were carried out

> with deliberation and slowly, and risks were avoided. This is characteristic of a regime which allows the individual little scope for initiative. From this, also, derives the practice of carefully rehearsing every attack, a procedure which becomes impracticable in the case of fast moving troops which have to advance into unknown territory. (Bayerlein 1956, p. 307–312)

The individual initiative of subordinate commanders was the exception rather than the rule. Before the war Isserson (2013, p. 12) had lamented that "operational art seems intolerably conservative." Being conservative was the safe option since most of the progressive Soviet military thinkers didn't survive Stalin's purges. War commenced with ill-qualified generals, but in the course of the war Soviets adjusted to the new situation and learned methods better suited to maneuver warfare. Some commanders, like Rokossovsky (1970, p 9), argued that "the main thing in my system of command personnel training was the rule to foster in each and every officer an aptitude for independent, resolute and bold action." This may be a sugarcoated impression of Soviet freedom of action as on one hand "the Red Army has never been an organization that favored imaginative, aggressive young leaders, whatever its rhetoric may say" (Leonhard 1991, p. 53). On the other hand, on the higher operational and strategic levels the Russian and Soviet officers have always been encouraged to be imaginative. It is on lower levels that rigidity and conformity overpower imagination and initiative when not mainly directed at better accomplishment of assigned tasks (Dupuy 1987, p. 40). Neither the Soviets nor Montgomery resorted to boldness, because of their superiority. Bold or timid, in the eyes of history the number of casualties doesn't really matter. Victories do. Therefore success is the determinant of the reputation of an operational artist.

> How much of this is really earned is extraordinarily difficult to determine. Even the best man fails against the irresistible power of circumstances, and even the average man must endure this power. Nevertheless, in the long run only the intelligent have good luck. (von Moltke 1993, p. 46)

Luck has an important role in the practical side operational art. This is because so much of the canvas of art of war is shrouded in the fog of war. Commanders make their decisions, win or lose, based on their limited grasp of the situation, their audacity, and their intelligence. Sometimes luck favors them and sometimes the decisions turn out to be wrong ones. Even after a prolonged period of lost battles and misevaluations when a commander becomes victorious again "we are always inclined to attribute the renewal of success to accurate calculations and clear insight, rather than to another turn in the wheel of fortune" (von Clausewitz, cited in von Freytag-Loringhoven 1991, p. 197). Luck is a necessary trait in any operational artist, but in the long run luck favors the genius because his risks are based on rational calculations. Luck alone is extinguished quickly. Skill and luck go together and no operational artist "should believe so fully in his star that he abandons himself to it blindly. If you are lucky and trust in luck alone, even your success requires you to the defensive; if you are unlucky, you are already there" (Frederick 1985, p. 394).

Successful gamblers don't make their living by relying on luck. They either cheat their opponents, which is a bad move at a card table but an acceptable stratagem in war, or use mathematics to increase their chance of winning. This doesn't mean reliance on complex theories but simple calculations of probabilities. Like a gambler, the operational artist needs to base his "decision upon the confusing mass of incoming intelligence must, generally speaking, be guided by the law of probability, and in the case of the enemy, too, he must assume rational action" (Goltz 1906, p. 127). Luck cannot be relied on, but a good deal of it is required for a plan to survive combat. Therefore, the operational artist must learn to manipulate his luck.

> Luck only stays with the good general who has a good system of command and control. (…) 'General's luck' surely comprehends three distinct through related elements—the creation of opportunity, the spotting of opportunity, and the exploitation of opportunity. Only in the second of these does pure chance, the unpredictable whim of destiny, play a part. (Simpkin 1985, p. 205)

An operational artist is always on the lookout for a 'lucky' moment. Luck is the ability to identify an opportune *kairos*-moment and to exploit its possibilities to the fullest. Occasionally in all gambling, be it poker or warfare, probabilities need to be cast away and one must rely on his intuition and psychological skill in comprehending the situation. War is a playground of the unexpected where strange things occur constantly. In an uncertain situation, "an inflexible clinging to intrinsic probability and persistent disregard of negative indications is apt to lead back to preconceived ideas, and may be the cause of fatal errors. How often does not the improbable occur in war!" (Goltz 1906, p. 127). Suvorov was an example of a wise gambler, because "battles excited Suvorov but did not carry him away. If he was more prepared to gamble than others were, he still knew when to stop." (Longworth 1965, p. 59). Warfare is fickle and

> great generals have often been beaten by inferior ones; but an exception does not make a rule. An order misunderstood, a fortuitous event, may throw into the hands of the enemy all the chances of success which a skillful general had prepared for himself by his maneuvers. (de Jomini 2007, p. 31)

Luck may give victory to the weaker. A commander may not waste time waiting for luck to favor him nor can he can disregard it. Simpkin (1985, p. 198) wrote about "luck management" and its call "for awareness and flexibility—the one a fact of the art of generalship, the other a product of directive control." The chances of being lucky are increased by being alert and active. Since there are special *kairos*-moments in battles, acting rapidly when one occurs is part of luck management just as well as using *Auftragstaktik* to create the same opportunities of direct action to subordinate commanders. Daring to act at these moments is making the best of the opportunity luck offers. As de Jomini (2007, p. 95) wrote, "a proper calculation of time and distances, joined to great activity, may lead to the success of many adventures which may seem very imprudent."

The bold gambler must on occasion step aside and be replaced by the calculator of probabilities. The art of command in war depends on the ability to judge the requirements of the moment and choose the correct stance to take. The operational artists need to use "less daring here, more daring there. Boldness that wins battles if it is risked also loses battles if it is not" (Delbrück 1997, p. 53). It requires flexibility of mind to assume contradictory modes of behavior depending on the situation. No matter how much the opposing forces differ in quality or quantity the outcome is never 'certain.' Machiavelli (1965, p. 202) wrote how "it is better to subdue an enemy by famine than by sword, for in battle, *fortuna* has often a much greater share than *virtú*." In warfare, it has always been necessary to try to eliminate the role of *fortuna* in the outcome and to rely on *virtú* (van Creveld 2008, p. 96–97). As Quintus (1988, p. 40) wrote of Alexander, "his sound strategy however, was shattered by fortune, which is more powerful than any calculation." The art of war entails so many interlinked factors that outcomes can only be estimated. There can be

> no absolute certainty in war, and yet it is not without some degree of relative certainty. We are comparatively certain about our own situation. We are very uncertain about the enemy's, but here too there are signs for us to read, clues to follow and sequences of phenomena to ponder. These form what we call a degree of relative certainty, which provides an objective basis for planning in war. (Mao 1963, p. 243)

While one must be willing to embrace uncertainty, it is still his task to do his utmost to lessen its impact. For von Clausewitz the art of command was to make decisions in the midst of uncertainty (Handel 2001, p. 8). "Only he will be successful, however, who dares to act in the face of the unknown. The future will be more lenient in judging the active than the inactive" (Guderian 1937, p. 39). In the most confused circumstances it is always better to take action and shape the situation, than to remain passive and float around on the waters of uncertainty allowing events and the enemy to propel one onward. Active participation in unfolding events mostly concerns tactical commanders. For operational artists and strategists the situation is different. The higher the level of war, the more time there is to decide and the more important the decisions become.

> It takes more strength of will to make an important decision in strategy than in tactics. In the latter, one is carried away by the pressure of the moment, caught up in a maelstrom where resistance would be fatal, and, suppressing incipient scruples, one presses boldly on. In strategy, the pace is much slower. There is ample room for apprehensions, one's own and those of others; for objections and remonstrations and, in consequence, for premature regrets. In a tactical situation one is able to see at least half the problem with the naked eye, whereas in strategy everything has to be guessed at and presumed. Conviction is therefore weaker. Consequently most generals, when they ought to act, are paralyzed by unnecessary doubts. (von Clausewitz 1989, p. 178–179)

To have too much time at one's disposal is not necessarily a good thing. Many experienced combat leaders use their instinct in decision-making. If one is hard-pressed to make a decision in a situation at hand it may be easier. Since there is little time, there is no second-guessing. Almost instinctive action has its benefits, but requires accumulated experience to work. An inexperienced commander is likely not to comprehend the events in toto and to err. In operational art and strategy, following von Clausewitz, the situation is more complex. An operational commander may produce his decision rapidly, but since the pace of action is slower and he is removed from the eye of the hurricane of combat raging around him, he has more time to ponder upon alternative solutions and methods. This doesn't mean that the plans of the brightest minds necessarily succeed. This is what happened with the Schlieffen Plan. It was never a sound formula for victory, but a great gambit whose success was dependent on several lucky accidents. As these didn't occur, the plan failed (Ritter 1958, p. 66). The more important the upcoming operation is, the greater the stakes and the burden of responsibility. It is a part of human nature not to make decisions with serious consequences without allowing thoughts of failure to influence the decision. If doubt begins to gnaw at the mind of the operational commander, despite the ample time for decision-making, it is possible that a decision will not be made in time, or at all. At some point contemplation must end and the operation commence. Schwarzkopf described what goes on in the mind of the commander at that moment. Once he had given the orders to attack in Desert Storm,

> I felt as if I were standing at a craps table in some kind of dream—I'd bet my fortune, thrown the dice, and now watched as they tumbled through the air in slow motion onto the green felt. Nothing I could do would change the way they landed. (Schwarzkopf 1993, p. 476)

In any case, no matter how well planned and prepared, waging war resembles gambling. According to Leo VI the greatness of the general is not determined by his ability to plan ahead when there are no time constraints and he can use the time in a leisurely manner. The gifted general hatches his plans rapidly when he is pressed for time, or, in the case of a true genius, when the decision has to be taken instantaneously.

> The mark of a genuine general and one worthy of admiration lies in perceiving what has to be done at the moment of great emergency rather than the ability to make plans about such matters before the emergency. (Leo VI 2010, p. 603)

4.10 *Coup d'Oeil* as Mastery of Time

There are many similarities and simultaneously differences between Japanese and Chinese interpretations of the art of war. The ancient Japanese military thought focuses on explaining the "Way" or "*do*" of warfare on both the level of an individual and an army (Cleary 2005, p. 5–6). To thoroughly familiarize oneself with the Way is the method of being prepared for anything at any time. "When the time comes, there is no moment for reasoning [...] Above all the Way of the Samurai should be in being aware that you do not know what is going to happen next, and in querying every item day and night. Victory and defeat are matter of the temporary force of circumstances" (Tsunetomo n.d., p. 16). The warrior focuses on being prepared to die at every moment of his life and armed. Decisions should be made in the space of seven breaths because "*a warrior is a person who does things quickly*" (Tsunetomo n.d., p. 30). Supposedly by constant preparation the man becomes experienced enough not to be perplexed by any circumstances (Tsunetomo n.d., p. 36). The Way emphasizes the meaning of the present, not the past or the future. There is no need to worry about the future or attach oneself to the past since it cannot be brought back. "A man's whole life is a succession of moment after moment. If one fully understands the present moment, there will be nothing else to do, and nothing else to pursue. Live being true to the single purpose of the moment" (Tsunetomo n.d., p. 45). In Occidental thought what has come to signify a commander living "for the purpose of the moment" is the intangible *coup d'oeil.*

Coup d'oeil is important for both the future and the present. The simplest description means distinguishing "the point chosen, before a battle, in the enemy's position, and on which the chief force of the attack is directed" (von Bülow 2013, p. 52). This is a passive interpretation, since it argues for only prescribing the decisive point prior to combat. de Jomini emphasized taking the initiative and seizing the moments during battle. This is due to the presupposition that the practice of operational art consists in "throwing the masses upon the decisive points, to do this it will be necessary to take the initiative. The attacking party knows what he is doing and what he desires to do; he leads his masses to the point where he decides to strike. He who awaits the attack is everywhere anticipated" (de Jomini 1992, p. 72–73). If both sides attempt to seize the initiative and cause a surprise "two armies approach each other, each intending to make an unexpected attack upon the other. A collision ensues unexpected by both armies, since each finds the other where it does not anticipate a meeting" (de Jomini 1992, p. 207). The more one seeks the initiative against an enemy with a similar doctrine, the more probable it is that both sides, regardless of their planning, will face an unexpected situation. The ability to act decisively and almost instinctively remains vital. "It is commonly said that modern war is the most recondite of things, requiring experts. War, so long as man risks his skin in it, will always be a matter of instinct." (du Picq 1987, p. 138–139).

Operational artists must understand that a part of winning time in war is developing measures for decision-making, giving orders, and overseeing their execution so that not a second is wasted, since every tick and tock of the clock until action is initiated is time lost to the enemy. Or must they? von Clausewitz opposed this kind of thinking since he argued that "all great commanders have acted on instinct, and the fact that their instinct was always sound is partly the measure of their innate greatness and genius" (von Clausewitz 1989, p. 71). Not every general can be a great commander and possess the genius permitting him to act with instinct in the most suitable manner. The artiste of war, the truly great genius, may do what he wills. The rest have to attempt to do their best to manage time.

The dilemma between making decisions by instinct or making them after painstaking contemplation of all aspects of the situation is resolved by von Clausewitz, when he argued that "any given situation requires that probabilities be calculated in the light of circumstances, and the amount of time available for such calculation will depend on the pace with which operations are taking place" (von Clausewitz 1989, p. 85). The pace of operations or the rhythm of battle determine how thoroughly the operational artist should contemplate. If there is sufficient time, probabilities for foreseeable changes in the situation should be calculated. If time is of essence, rapid methods have to be employed. The more quickly one wants to bring about a resolution, the more planning, calculation, and contemplation should take place prior to execution. Delbrück (1990, p. 321) argued that in the art of war "not everything can be calculated, weighed and measured; in situations defying such calculation, the belief in his own star must govern the commander's decision." The time for estimation and calculation is before action.

For Frederick (1985, p. 346), "a perfect general, like Plato's republic, is a figment of the imagination." It would be an admirable goal for every operational artist, but humanity doesn't produce flawless individuals. The great general should be "more than an industrious, active, and indefatigable man, not forgetting one thing to execute another, and above all not despising those sorts of little details which pertain to great projects" (Frederick 1985, p. 346). These are characteristics of extraordinary people. Activity, indefatigability, and execution all relate to the time-factor. The mysterious concept of "*coup d'oeil*" can be found in the texts of many of the great Occidental military thinkers. But what does this 'stroke of the eye' mean and why is it listed so high in the rank of qualities a general needs? It can be translated as 'a glimpse,' since it refers to the idea that "a general who possesses a *coup d'oeil* is with a glimpse of an eye able to perceive the situation and act accordingly" (Frederick 1985, p. 341). Frederick claimed that when a general is accustomed to the size of his army, his *coup d'oeil* will develop accordingly and he can quickly grasp the ground he can cover with the troops at his disposal. He summarized the essence of *coup d'oeil* as "the talent which great men have of conceiving in a moment all the advantages of the terrain and the use that they can make of it with their army" (Frederick 1985, p. 341).

von Clausewitz (1989, p. 578) in turn interpreted the *coup d'oeil* as the "ability to see things simply, to identify the whole business of war completely with himself" and argued "that is the essence of good generalship." In this sense *coup d'oeil* is a somewhat inborn personality trait of a commander. de Jomini argued that theory is an uncertain guide and thus having all theoretical knowledge can never be as valuable as "a natural talent for war, nor be a sufficient substitute for that intuitive coup d'oeil imparted by experience in battles to a general of tried bravery and coolness" (de Jomini 1992, p. 203). Thus, *coup d'oeil* is a natural talent that still accumulates and grows with experience. It is crucial to an operational artist since, "only if the mind works in this comprehensive fashion can it achieve the freedom it needs to dominate events and not be dominated by them" (von Clausewitz 1989, p. 578). It is the nature of war to surprise and unless the operational artist can prepare himself to react instantaneously, the events will overwhelm him.

Frederick saw a special need for *coup d'oeil* in two tactical situations: when one encounters the enemy on his march and when one finds the enemy in position and must attack. In the first case the general must quickly choose the ground on which to fight and determine how to conduct the combat so that all advantages and disadvantages of the terrain are accounted for in the battle formation. The rigid tactics of his time provided models for orders of battle, and the instant decision concerned finding the most suitable one. In the second case the *coup d'oeil* supplies the general with the ability to perceive where he should direct the force of his attack. According to Frederick, "Whoever has the best *coup d'oeil* will perceive at first glance the weak spot of the enemy and attack him there" (Frederick 1985, p. 343). *Coup d'oeil* has gradually become more of a metaphysical concept that today refers mostly to the inner eye of imagination and intellect. This is partially an outcome of the enlarged battlespace.

> In the 18th century, the *coup d'oeil* of the leader had its place on the tactical battlefield even in the physical sense. The small armies and the accepted methods of fighting enabled him, as a rule, to watch not only his own troops but also the enemy's, which would be out of the question today for the commander of an army corps, and hardly possible for that of a division. (von Freytag-Loringhoven 1991, p. 285–286)

von Clausewitz noted that traditionally the idea of *coup d'oeil* was interpreted as a glimpse of the actual eye and turning the visual information of the battlefield into suitable modes of action. *Coup d'oeil* resulted in a calculation of time and space and evaluation of their interaction.

> Because time and space are important elements of the engagement,[…] the idea of rapid and accurate decision was first based on an evaluation of time and space, and consequently received a name which refers to visual estimates only. Many theorists of war have employed the term in that limited sense. But soon it was also used of any sound decision taken in the midst of action—such as recognizing the right point to attack, etc. *Coup d'oeil* therefore refers not alone to the physical but, more commonly to the inward eye. (von Clausewitz 1989, p. 102)

This tendency to use the inward eye was already evident in the second example of Frederick we discussed. The first one, figuring out a suitable order of battle is a perfect example of practical evaluation of time and space. Finding the weakest part of the enemy's defense is accomplished by the eye of the mind even more than the physical one. Seeing with the inward eye requires powers of intellect and imagination. Thus, to develop

> the power of imagination and its various ramifications is an essential part of general staff training, and an indispensable requisite for leaders of large forces distributed over a considerable area. The ability to form accurate mental pictures of a situation quickly is especially important today when the

> higher commander cannot hope to see his troops with his physical eyes. (von Freytag-Loringhoven 1991, p. 296)

General staff education tends to develop other traits than imagination, but only through ability to be imaginative the complexities of battlespace or theater can be comprehended. They are no longer under the commander's direct observation by the physical eye in real time. Thus, the inner eye must add detail to the rather sketchy view provided by different channels of information. *Coup d'oeil* as the ability to manage available force in the optimal relation of time and space is thus despite its instantaneous essence a product of the intellect and not a reflex. This is why von Clausewitz extrapolated from *coup d'oeil* the concept of "*presence of mind*" which, according to him,

> must play a great role in war, the domain of the unexpected, since it is nothing but an increased capacity of dealing with the unexpected. We admire presence of mind in an apt repartee, as we admire quick thinking in the face of danger. Neither needs to be exceptional, so long as it meets the situation. (von Clausewitz 1989, p. 103)

One inherent characteristic of an operational artist is the skill to lose no time in evaluating the situation from all angles prior to making a quick decision how to act. It has been argued that courage is the primary requirement of a soldier (von Clausewitz 1989, p. 101). For an operational commander, the ability to use *coup d'oeil* is the way to show his courage. "Determination in a single instant is an expression of courage [...] we are referring not to physical courage but to the courage to accept responsibility, courage in the face of a moral danger" (von Clausewitz 1989, p. 102). When making a decision the operational artist is aware he will face the consequences but still possesses the courage to decide.

Especially at operational level of command, decisions should be based not on "the physical situation as last reported or observed, but the commander's mental picture of the situation as it will shortly develop" (Simpkin 1985, p. 224). Thus *coup d'oeil* is concerned with looking into the past, present, and future at the same time. When the operational artist uses it to look into the past, he plots the causal chain of events that have led to the present. The present consist of the battlespace he sees with his own eyes but also through all reports on the situation he receives. This is the easiest part of employing the *coup d'oeil* and hence demands increase.

The insufficient information of the aforesaid reports are used to produce a vision of how the situation will evolve and to plot the course of future actions. Through imagining short-term future developments the commander wins time by mentally preparing himself and his subordinates to emergent situations.

Much of the mastery of *coup d'oeil* is concerned using the mind's eye to pierce the fog of war to augur what will happen. Nevertheless, we should not think that the commanders of the old age who were able to view the battlefield physically in front of them had things easy either. Even if Caesar, Hannibal, and Scipio, or later Napoleon and Frederick the Great, enjoyed this advantage, they required the inner eye as well since

> the annihilating battles fought by these great generals were possible only because they possessed in high degree not only the outer but the inner *coup d'oeil*—because they possessed that power of determination which is always an inseparable part of battle intuition and which guided them not only during but after the battle as well. (von Freytag-Loringhoven 1991, p. 286)

Ludendorff claimed that in a war of maneuver the commander is required to form and reform mental pictures of the prevailing conditions in colorful sequences. His decisions must at any given moment be based on his own capacity and *coup d'oeil.* In war of movement "the handwork of the soldier becomes an art and he is turned into a commander" (Ludendorff 1919, p. 37). This is due to the need to make individual decisions on the spur of the moment at any given time. Since movement is quick, conditions change rapidly. Intuition, imagination, and intellect are necessities both during the battle and after it. To plan a future course of action the operational artist must preconceive how the outcome of battle influences enemy dispositions. The chaotic circumstances of the battlespace serve to confuse. The operational commander's inner *coup d'oeil* must gauge the nature of the situation and proceed to fulfill his operational idea and continue the operation without delay while the enemy is still physically and mentally disorganized.

While we have thus far concentrated on directing battles and operations, *coup d'oeil* as a leadership skill must "also have its place in strategy, since here as well quick decisions are often needed" (von Clausewitz 1989, p. 102). According to Mao *coup d'oeil* is first and foremost a process of employing one's intellectual powers to discern the situation as a whole and not focus on detail. To make sense of the strategic picture, mental powers have primacy since because the entire situation and context are not visible to the physical eye. The same need for abstract contemplation applies to operational level *coup d'oeil*, because

> what pertains to the situation as a whole is not visible to the eye, and we can understand it only by hard thinking; there is no other way. But because the situation as a whole is made up of parts, people with experience of the parts, experience of campaigns and tactics, can understand matters of a higher order provided they are willing to think hard. (Mao 1963, p. 81)

The proper utilization of *coup d'oeil* requires contemplation of the situation on at least one level higher than one's own. That is, the tactical commanders must relate the battle in its operational context to perceive what is required. Tactical victories or defeats shape the course of operations, and operational artists must conceive of their activities within the context of strategic objectives. Even the commander-in-chief has to comprehend the political situation to support his perception of opportune moments when to alter his strategic plans to best suit the objectives of the state. Many theorists don't use the specific term *coup d'oeil*, but the idea is illustrated in most of the texts in varied terminology. Andolenko appraised Suvorov for his "*quick grasp*", defining it as

> the power of solving, swiftly and well, any kind of problem likely to arise; of appraising a situation quickly, of making a decision rapidly; of preparing its execution rapidly, yet with attention to detail; then, finally, of carrying it out in the same way, with the maximum chances of success. Reason and calculation, an accurate knowledge of the possibilities of one's own side and the enemy's, play an important part. (Andolenko 1956, p. 15–16)

Every important characteristic attributed to the *coup d'oeil* of the great captains is present in this quotation. Even the occasionally omitted ability to combine reason and calculation with rapidity of decisions and executions is included. As another example we can use de Saxe who wrote that the truly great captain needs

> a talent for sudden and appropriate improvisation. He should be able to penetrate the minds of other men, while remaining impenetrable himself. He should be endowed with the capacity of being prepared for everything, with activity accompanied by judgement, with skill to make a proper decision on all occasions, and with exactness of discernment. (de Saxe 1987, p. 294)

How could we better summarize the necessary characteristic of an operational artist? Here we see the importance of planning and preparation and the need to focus on the intentions of the enemy to prepare for them combined with the ability for instinctive and improvised action. de Saxe saw that art of war consisted of two branches. One,

> that is to say drill and the method of fighting, is methodical; the other is intellectual. For the conduct of the latter it is essential that ordinary men should not be chosen. Unless a man is born with talent for war, he will never be other than a mediocre general. It is the same with all talents; in painting, or in music, or in poetry, talent must be inherent for excellence. (de Saxe 1987, p. 296–297)

This division of the art of war into the methodical and the intellectual parts is close to Musashi's (1995, p. 47) idea of the art of war being two-fold, that of the sword and that of the pen. Time and its management are represented in both spheres. The methodical works on *kronos*-time and seeks to save time for commanders from mechanical responses by perfecting doctrines to shorten the response time to orders or enemy action. The intellectual and inspired side of the art of war—the planning, the *coup d'oeil*, the initiative, the surprise factor, and the command and control function—attempts to hasten its processes too but depends on exploitation of the *kairos*-moments. A true genius might not need additional time because of his "talent for sudden and appropriate improvisation" (de Saxe 1987, p. 294). Sadly, most operational artists are ordinary men facing extraordinary situations and to survive let alone master them, every second spared from manual drafting of orders and reallocated to thinking them through is important. Armed forces treat time as a resource to be distributed where it is needed. The management of time is the task of the operational artist who must know how to

> subsist his army and how to husband it; how to place it so that he will not be forced to fight except when he chooses; how to form his troops in an infinity of different dispositions; how to profit from that favourable moment which occurs in all battles and which decides their success. All these things are of immense importance and are as varied as the situations and dispositions which produce them. (de Saxe 1987, p. 295)

This summarizes de Saxe's thoughts on time and its meaning. While the tasks of the commander are chimerical, he needs to manage time and temporize and synchronize all the activity of the troops. Most importantly, he must spot the auspicious *kairos*-moments amidst the humdrum of war. The operational artist thrives on these special *kairos*-moments, and *coup d'oeil* is the personal trait that enables their identification. In contemporary warfare passively waiting for the right time is no longer enough and the commanders need to actively manage their luck and proactively create *kairos*-moments for exploitation. During the Third Wave both operational art and

> strategy will require not just the creation of strategic options, but the rapid creation of them—at a pace faster than the enemy can match. Warfare is and will remain a time-competitive event. (Leonhard 1998, p. 157)

This chapter focused on the mental and intellectual aspects involved in the management and control of time and temporality. Commanders of today cannot manage all issues involved in operations. Staffs were created to support the commander by managing the details and allowing him to practice operational art. In addition, certain methods of command like *Auftragstaktik* were invented to rationalize the command and decision-making processes. To save time and to reallocate it into a more important activity, the chain of command should be as short as possible and authority to make decisions delegated to those subordinate commanders who are directly involved in the situation. Time is saved by not having to ask authorization but by awareness of the commander's intent and acting accordingly.

While *Auftragstaktik* considerably shortened the chain of command, it was also recognized that the chain should be as flexible as possible. Toffler identified "flexi-time" as one of the characteristics of the Third Wave and this chapter argued that to adhere to flexible temporality the command of armed forces needs to be just as flexible. To manage the complexity of warfare every process and product should be as simplified as possible. Operational flexibility is a requirement for the minds of the planners, the commanders, and the resulting plans. The mind, the plan, the decision-making, the command, and the execution—all of these have to exhibit plasticity. Only then can fluid operational art be created in contrast to a rigid one strictly defined by doctrine and procedures.

The mechanical way of winning time intellectually is to hone the processes of information gathering, analysis, and decision-making so that time can be saved from secondary functions and reallocated to directly support the commander. Time lags in the processes must be minimized to ensure the commander can receive sufficient information in time. Too little information doesn't disperse the for of war and too much information only serves to confuse. The information age has to rethink how much information is actually needed, because refining information into knowledge takes time and even operational artists don't have to be omniscient.

Nevertheless, there are no certainties in warfare, and with every decision, risks are taken. Occasionally commanders resort to gambling, but mostly they manage the risks and evaluate them in relation to possible outcomes. The commander's most important means of managing temporality is to make informed decisions as rapidly as possible. This depends on his *coup d'oeil* as a combination of the physical and the inward mental eye in creating situational awareness through imagination and intellect alike. Energetic, intelligent, and imaginative operational artists, able to combine these three personal traits, are the masters of time in warfare.

4.11 Time to Think, or Conclusions

Development of military thought is not teleological and progressively improving at a set pace toward perfection. Evolution in operational art moves in waves and cycles. Old ideas return clothed anew and sometimes slightly altered to suit better the new context. We tend to

> rediscover that which was never lost, but was only misplaced and forgotten and which probably should never have been forgotten in the first place … strategic ideas rise and fall, appear and disappear, and then reappear in slightly different form, according to the policy and strategic agenda. (Gray 2007, p. 62–63)

Ideas come and go but some things remain constant even if the meanings given to them change. Time is one of the basic elements of warfare that has to be conjoined with force and place to create operational art. The commanders throughout the ages have employed time and tempo to coordinate their operations and combined many factors into a unified whole according to a pre-planned timetable. To do this effectively the operational artist must comprehend both the universal and 'wave-specific' aspects of management of time. The operational artist requires a Janus-like approach to his art. One face needs to be turned to the past and the other boldly facing the future.

> The aim of military study should be to maintain a close watch upon the latest technical, scientific, and political developments, fortified by a sure grasp of the eternal principles upon which the great captains have based their contemporary methods, and inspired by a desire to be ahead of any rival army in securing options on the future. The ways in which Napoleon achieved strategic surprise are of little guidance nowadays, but his ruling idea of strategic surprise ought to dominate the mind of any commander. The practical value of history is to throw the film of the past through the material projector of the present on to the screen of the future. (Liddell Hart 1927, p. 173)

Lessons of the past don't work verbatim in the world of today. Past victories cannot be recreated down to the last detail, because "history knows of no two wars, no matter how close together, that were conducted by the same methods" (Sokolovsky 1963, p. 237). Thus, the aspiring operational artist has to skip over the case-specific elements of the character of warfare and grasp the universal ones. He must understand the principles previously used to create desired outcomes, adopt and adapt them to the demands of the present, and through his intelligence and imagination perceive their role in the future. Operational artists should look into both the past and the future for inspiration. The technical, scientific, and political development trends set the course for the future. The soldier must anticipate it and ride the crest of the wave of progress to retain the ability to use the age-old principles in new contexts.

von Moltke (1993, p. 24–25) argued that "human life, even the entirety of human nature, is nothing but war of the future against the present. The lives of the various peoples are no different." The biggest conflict occurs when ideas of tomorrow clash with the practices and ways of thought characteristic to today. Society and the wars it wages evolve constantly but cyclically. History doesn't repeat itself but we benefit from looking at the lessons of history to help us overcome the challenges of tomorrow. Unless we distinguish between the valid and the obsolete teachings of history and re-evaluate the principles of war, we are liable to repeat past mistakes by basing our actions on wrong lessons and thus forcing historical patterns to reoccur.

None of the great theorists, not even von Clausewitz, should be taken as sacrosanct but instead constantly challenged. Not all of his thoughts remain valid and there are considerable gaps and inconsistencies in his text (Echevarria 2013, p. ix–2). Strategy and operational art need to treat 'the classics' in a somber fashion. For some they have become dogma not to be criticized. Others in their frustration become iconoclasts and seek to prove the uselessness of classical theories in the Third Wave context. Neither of these vantage points should be adopted. When visionaries err, they do it with the same enthusiasm and boldness as when their thoughts triumph. The age of mechanization and the indust-reality in general was reflected in theories of operational art as one of vigorous offensives and maneuver and it is hard to recognize this in Fuller's prediction that, "warfare is likely to become less offensive once the horde is diminished, or disappears, consequently wars will become less frequent" (Fuller 1932, p. 280). The reality of the twentieth century turned into something completely different. There is no guarantee the twenty-first century developments will follow any predicted pattern either.

Military minds of past and present have read their von Clausewitz, Sun Tzu, and de Jomini and accept them as 'holy writ.' The older the classical text, the more its ideas have been copied by later thinkers. Since it is a well-recognized principle that the fundamentals of strategy do no change over time, we are confronted with recycled thoughts inherited from one generation to another without critical evaluation. Nobody argues that striking at the right time at the right place with the right force is not the key to victory. But this has been repeated so often that it has become an empty cliché devoid of any meaning.

We need operational artists who can give us guidance on *how* to determine the right time and place. Due to the complexity of war in our age, comprehensive lists are impossible to produce, but some guidelines are needed lest we degenerate to the level of parrots repeating hallowed slogans that passing of time has hollowed. The importance of principles has not waned but their meaning has to be reinterpreted anew from time to time. To pick an example, the idea of concentration of force has turned into concentration of effects and it can be accomplished differently with precision weapons than with armed peasants or tank armies. Yet the difference is superficial because regardless of the tools used in exerting force, the necessity to do so remains the crux of operational art.

This idea applies to the meaning of time just as well. In warfare of different waves there are different amounts of time to handle, different ways of winning and saving it, but time is always of essence and somehow or another, it needs to be managed and optimally utilized to support operations. The idea of quantifiable time imbued with the purpose of saving every second characteristic to indust-reality is not wholly compatible with the idea of flexi-time that has to be considered as a part of the Third Wave way of war. Time management will still grow in importance with the constant acceleration of the tempo of warfare. The principles of war must be analyzed to determine *how* they influence operational art specific to a certain wave of warfare and how they should be utilized in time management.

In time, even some of the most basic concepts of warfare experience change. von Clausewitz (1989, p. 109) wrote that "*one cannot conceive of a regular army operating except in a definite space.*" The definition of space must be rethought to fit this argument into today's context. Time and space were central to Clausewitzian thinking and their relationship is a dialectic one on all three levels of warfare (Strachan 2007, p. 41–42). Tactically and operationally in Clausewitz's time this definite 'space' meant a two-dimensional battlefield, consisting of only depth and width. The invention of aircraft added the third dimension of height and created a battlespace. The German V-2 rockets were a developmental forerunner of many of the precision weapon technologies of today and there were predictions that "whatever form the future may take, the rocket will certainly play a very large part in every form of warfare" (Martel 1945, p. 72). The rocket pushed the boundaries of battlespace further because

> it adds a new sphere of movement to those existing: movement in a vacuum. This possibility is as great if not a greater revolution than that introduced by the aeroplane, because it raises war into pure space. (Fuller 1948, p. 320)

When ballistic missiles and spaceflight became possible the potential extent of battlespace has expanded far beyond the surface of the Earth. Simultaneously the ability to manipulate the electromagnetic spectrum has opened a parallel and virtual dimension of battlespace, which has since been widened to the intangible cyberspace. How can we argue that any contemporary armed force could conceivably have a 'definite space' to operate in? The answer is simple: time frames the battlespace. An operation has to have temporal boundaries, and warfare always takes place in a spatio-temporal locus. If battlespace has extended beyond comprehension, the idea that warfare starts is carried out and ultimately ends at a certain point in time keeps it manageable. War can be waged almost anywhere at any time but not everywhere all the time. Time sets boundaries and creates a frame for fluid operations to flow in. This is characteristic to the Third Wave warfare with its fuzzy temporality and flexible management of time. For the classics of the past war began and ended at a definite point along the line of *kronos*-time. Today the temporal boundaries have to be set artificially, if necessary.

Time is important in warfare on all levels but has different implications and requires different management. Time alone doesn't bring about victory or defeat, but its manipulation in terms of mobility, surprise attacks, and catching the enemy off his guard perhaps will. In a similar manner the terrain is inconsequential, but how one employs it in support of operations provides the keys to victory. Numerical superiority of troops doesn't amount to anything, unless one can use them optimally to attain operational and strategic objectives. We return to the classic military maxim of using the right amount of force in the right place at the right time. von Clausewitz (1989, p. 207) wrote that combat itself "is the essential activity of war, but we must also consider men, time and space, which are the components of this activity." His approach was battle-centric, and after the war started, he saw fighting as the only means of winning it (Echevarria 2013, p. 3). Combat, the actual *praxis* of warfare, is a sum of different components. The crucial factors of time, place, and force are worthless by themselves. They gain importance only when combined by the general's mental and physical activity and in relation to the enemy. Only through action—or inaction—at any given place and time, the importance emerges. Every moment of an operation has the potential to be the most important one, but only if action is undertaken or discarded at that time. Only when this trinity is synchronized harmoniously, we may expect a masterpiece of operational art to unfold. The roof of the Sistine Chapel is covered with basic colors of different paints, but Michelangelo created something far greater than the sum of ingredients on his palette. The operational artist like any other artist must apply his genius to his tools. Time is a tool, a resource, and a precondition of his art.

The need to manage, manipulate, win, and save time takes different forms in each age. When time is not only measured and saved in one's own actions but actually 'robbed' from the enemy, we are no longer talking about *kronos*-time measured in minutes and seconds but 'relative' time. The meaning of time in operational art is tied to what can be accomplished within a given time-period, and the faster one is, the less time enemy has at his disposal. Thus time is relative not only to what one can accomplish during it, but also to what the enemy can do and how 'robbing' his time can hamper his operations. Time won or wasted in the heat of combat or over the course of the operation has meaning through the influence it has on enemy and his temporizing.

Emphasis on winning time has led us to remain true to the principles of indust-reality and to create streamlined processes of maximized efficiency. To utilize time and to manipulate it as a resource is crucial, but if we focus entirely on 'winning seconds' and seeing the clock as our principal enemy, we quantify warfare. Warfare turns into a test-lab of natural science instead of the intuitive playground of imaginative art. Quantitative and qualitative factors must be combined in contemporary operational art. If we produce avant-garde art, there is a looming possibility that the result is too creative to function. If we seek to compute warfare, the importance of intuition, imagination, and spontaneity disappears. Commanders are no longer artists of war but its engineers. Yet there is a long tradition to perceive war not only as an art, but a science as well. "The essential of tactics is: the science of making men fight with maximum energy. This alone can give an organization with which to fight fear. This has always been true. We must start here and figure mathematically. Mathematics is the dominant science in war, just as battle is its only purpose" (du Picq 1987, p. 166). This doesn't exclude creativity and intuition or make war less of an art. Mathematics is the basis of all arts: the musical notes adhere to mathematic equations and laws, perspective is dictated by mathematics, and proportions follow mathematical rules. Mathematics is important in operational art as the foundation of the masterpiece and not its method.

Rhythm of warfare is another concept which gains its meaning only in conjunction with timing. A symphony orchestra would sound cacophonic if the notes each individual instrument produces were not perfectly timed in relation to the others. The operations of armed forces must be similarly synchronized, that is, timed with every element in joint relation to others. Operational art can produce a beautiful symphony and avoid the fugue created by the fog of war. The commander becomes a conductor if his intuition of timing his operations is flawless. When the instruments he marshals begin to produce discord, he should be able to 'play by the ear' and resynchronize them again.

When the commander-in-chief of the army was simultaneously the head of state, grand strategy, policy, and strategy were one. When there are no restraints on the commander-in-chief, he is able to make decisions and carry them out so rapidly that the enemy is hard-pressed to keep up with the pace. This concentration of power would be inconceivable in the democratic societies of today, and even in the past it existed only as a historical anomaly. "Subordinating the political point of view to the military would be absurd, for it is policy that has created war. Policy is the guiding intelligence and war only the instrument, not vice versa" (von Clausewitz 1989, p. 607). No matter how much more effectively war could be waged and how much time won without the restraining leash of policy, political control alone can ensure that war will not escalate beyond reason. Political goals restrain the militaries from deciding the ultimate conclusion of the clash of arms by the most destructive means at their disposal.

> So policy converts the overwhelmingly destructive element of war into a mere instrument. It changes the terrible battle-sword that a man needs both hands and his entire strength to wield, and with which he strikes home once and no more, into a light, handy rapier—sometimes just a foil for the exchange of thrusts, feints and parries. (von Clausewitz 1989, p. 606)

And the world needs this foil, because technology has created weapons that make warfare potentially so deadly that combat must be tamed into a political argument "which takes up the sword in place of a pen" (von Clausewitz 1989, p. 610). War is another form of political discussion beyond exchange of diplomatic notes, "Its grammar, indeed, may be its own, but not its logic" (von Clausewitz 1989, p. 606). This discussion may be carried out in harsh tones. When forced to wage war, everything should be done to win time from the enemy and end the war. But policy has to decide which means of winning time are allowed. This requires expertise, insight, and decision-making capability from the political leadership of the nation.

> Woe to the government, which, relying on half-hearted politics and a shackled military policy, meets a foe who, like the untamed elements, knows no law other than his own power! Any defect of action and effort will turn to the advantage of the enemy, and it will not be easy to change from a fencer's position to that of a wrestler. A slight blow may then often be enough to cause a total collapse. (von Clausewitz 1989, p. 219)

This creates a situation where the politicians need to be thoroughly familiar with their instrument so that the operational commanders can focus on perfecting their art so that the military loses no time in reacting to political orders, issuing its own, and carrying them out. Since war is an expression of political will of the society, it must follow the written and unwritten rules of that society. If the pace of living in a society has quickened, it is self-evident that warfare adapts to this pace. But generals are rarely avid futurists, and the intricacies of the change in tempo are difficult to analyze when one is confronted with future shock. How can we attempt to augur the future if we struggle to comprehend the present?

We could do worse that dig deep into the traditions of operational art and strategy for guidance. Why are these old texts written about wars totally different than the ones we are likely to face in the future still necessary reading for future officers? The wisdom in the best texts enables them to rise above their contemporary context and treat war as a universal phenomenon. We can even find predictions of things to come.

> Turning back to the law of military development, and remembering that the present tendency of civil science is towards the existence of an electrically constituted universe, and that industry and civil life are becoming daily more influenced by electricity, and the many applications of this energy, it is conclusive that military organization will follow suit, and will develop what I will call, for want of a better name, the 'robot' cycle. (Fuller 1932, p. 229)

Fuller wrote these words over eight decades ago when "an electrically constituted universe" was an absurd idea. Today, with internet, wireless connections, cloud computing, networks, cyberspace, and virtual reality, Fuller's prediction seems to be accurate. We might not yet live in a "'robot' cycle," but such a phase has potential to emerge. During immersion in the thoughts of past theorists and practitioners, it is beneficial to augment these ideas with the dreamers of today and their visions of what the future could be. New technologies enable wars to be fought with new means but they don't alter the essence or nature of war. War abides to certain principles but the application of these principles varies according to not only technological but also societal and cultural level of development. In the past, the present, and the future wars will be fought for political purposes.

All three waves of warfare coexist today, and operational art should utilize them in unorthodox ways. For example, asymmetric warfare has always been present. It is the oldest element of war from the time when a caveman had to use a weapon as defense against dangerous predators. All prominent and superior forms of warfare from the *grande armée* to mechanization and drones have led to a resurgence of the guerrilla tactics, the most common of all asymmetric forms of fighting. As Fuller wrote, "guerrilla warfare, the most primitive of all forms of war, is likely to be revived, and as it obviously demands a high order of initiative to combat it, it will force this essential quality upon the commanders of organized forces" (Fuller 1943, p 11).

It has further evolved during the Third Wave as an asymmetric response to superior forces in almost every sense. The juxtaposition of speedy modern maneuver warfare and its slow asymmetric counterpart is characteristic to many recent conflicts. The growth and spreading of asymmetric warfare have influenced the Third Wave thus far. Asymmetry today works not only through variable tactical choices or variable pace of activity but is an increasingly important element in all facets of operational art because it can employ variable timing. If, as this book has repeatedly argued, warfare has become too complex and its pace has accelerated beyond control, measures to counter these developments have to be invented, and asymmetric methods look promising. They offer the possibility to break free from the spin of the decision-making cycle and self-sustaining processes that erode the artistic element of war and favor the mechanistic parts. An asymmetry of time is an element of warfare endemic to the Third Wave.

First and foremost, asymmetry is an option that needs to be considered as a means of controlling, managing, and optimizing time in warfare. Asymmetry promises the option of finding the auspicious *kairos*-moments of when to act to derail the processes of the enemy. The Third Wave warfare should combine the two earlier paradigms and add new characteristics particular to itself. There has to be a return to the agrarian focus on the importance of *kairos*-moments and their uniquely temporary nature while a sense of efficient management of quantifiable *kronos*-time has to be included as a carry-over from indust-reality. Third Wave warfare needs to return to identifying the natural rhythm of activities and seizing the moments but simultaneously attempt to reset the rhythms to create more potential *kairo*s-moments for exploitation. In highlighting the importance of moments of *kairos*-time in the flow of *kronos*-time, the Third Wave society with its emphasis of flexi-time pays homage to the agrarian society without discarding the importance of measurable time indust-reality embraced. Temporal resources during the flow of *kronos*-time have to be meticulously managed and not a second can be wasted, but when a moment of *kairos* arrives, intuition, imagination, intellect, and inspiration are required to fully exploit its short but auspicious duration. The essence of managing time in operational art in the Third Wave societies must be the ability to configure and assimilate the ideas of temporality from the two previous Waves as well.

> An important requirement in military art is a skilful combination of styles of warfare that will respond properly to the concrete situation of a given place and time. Each style of warfare must be adapted to the balance of forces between the enemy and ourselves and to the strategic situation of each phase of war. (Giap 1970, p. 89)

To summarize what a future war will be like, it is a safe bet that whatever new tools and tactics will be employed, the essence of war, its raw and violent nature, will remain unchanged. If we look back at the development of warfare during the last decades we note that it is not the drone or any other Third Wave technology-enabled weapon that has produced most casualties, but the unholy trinity of the machete, the AK-47, and the suicide bomber (Smith 2008, p. 299–300). It is unlikely that the next wave will be an irresistible tsunami that would wash over warfare altering its foundations. Rather warfare is likely to become multi-spectrum, or 'multi-wave', war in which the range of operational art stretches from the principles governing agrarian age, the industrial age, and the information age alike. All three waves and their ways of thought and action will be manifested in the asymmetric battlespaces of tomorrow. As Marcus Aurelius (2002, p. 78) wrote, "Don't fear the future. You will face it, if that is your fate, armed with the same reason that protects and guides you in the present." Unfortunately the military cannot stand back and stoically follow the development but must strive to adapt.

The art of war and operational art must adapt to the social, political, and civilizational contexts in which they would be employed and not attempt to mold their external environment to adapt to them. As Stempel (2012, p. 169) wrote, "as humanity continues to develop, war alters to keep pace; that has been war's great constant since its inception." The ways and means of fighting are largely dependent on existing technologies and cultural patterns of thought, but wars will be waged for political purposes. Their form will change, their nature will not. A good example of how to adapt and to avoid future shock comes from General von Kleist who in his time was faced with a new invention: "Very well, then, let us pass without hesitation into the age of smokeless powder. The world still belongs to the bold" (Cited in von Freytag-Loringhoven 1991, p. 276).

All in all, managing time in operational art can be considered to be an artform on its own right. While the 'need for speed' seems to be integral part of military thought process, there exists an even more urgent need to control the speed. On occasion slowing down, pausing, or perhaps restarting the operation with an adjusted tempo and better synchronization of all elements involved is the most prudent way of managing time. The military mind is infatuated with bold audacity and dashing maneuver because in mobile warfare they have traditionally brought about decisive victories along with numerous operational blunders we rarely read about in history of war. Operational art is not only about painting pictures of martial glory or enacting spectacular performances on the theater of war. To combine tactical battles into operations and create operational art requires orchestration of all the tools at the disposal of the commander and synchronization of their employment and effects.

While the artistry in war manifests itself slightly differently on all levels of warfare, especially operational art is about finding and creating rhythm and tempo for warfare. Synchronization of effects requires more than theatrical excellence or just timing the activity of the troops. It predisposes a rhythm composed of harmonious melodies flexibly combined with wolf tones. At one moment during an operation the flowing liquidity of Vivaldi has to make room for grandiose and massive use of instrument in the booming forceful notes of Sibelius and vice versa. Rhythm is of utmost interest in operational art but it has to incorporate sudden shifts and changes enabled by a profound understanding that there is no universal rhythm of war. The rhythm, or the pulse, of warfare varies depending on the type of war waged, the desired outcome of the operation, and especially in relation to all the unexpected and unforeseen that not only can but inevitably will occur within the battlespace and beyond it. Often the most recent developments in the art of war prove insufficient in pursuit of operational or strategical victories. The tested and true has to be combined with the unorthodox and innovative. The mechanical factor enabled by the latest technological gadgets has to conjoin with the artistic and imaginative aspect only human intuition can provide. Since war is the realm of the unexpected, the operational artist should strive to increase the number of unexpectable element for his enemy. As times have changed the outlook of war has changed while its essence has remained immutable. The ontological nature or war remains the same as ever, but in each time and era the character of warfare has progressed simultaneously with the societies waging war. Operational art thus is very much a product of its time. It is the practical employment of art of war in changing contexts, and the commander must remain rooted to the present with his gaze fixed in the future while understanding and preparing to employ the methods of past eras in new ways to face the challenge of bold new times.

The management of time in operational art requires making the most of the time at the commander's disposal but simultaneously recognizing and exploiting the auspicious *kairos*-moments in the flow of time. The traditional strong-willed energetic commanders are as important for execution of operations as the original and imaginative thinkers who provide the artistic element and introduce a rhythm in their planning phase. To occasionally discard the watches to embrace time itself is a prerequisite of operational art. While the mechanized indust-real warfare ticked along like a clockwork the rhythm of contemporary operations cannot be set with a metronome. It requires an inner ear for rhythm. Thus an operational artist must employ not only the *coup d'oeil* but *coup d'oreille* to deliver his *coup de grâce*.

References

Alberts, D. S., Garstka, J. J., & Stein, F. P. (2000). *Network Centric Warfare – Developing and Leveraging Information Superiority* (2nd ed. (revised)). Washington, D.C.: DoD C4ISR Cooperative Research Program.

Alexander of Tunis, Field Marshal Earl. (1962). *The Alexander Memoirs 1940–1945*. London: Cassell.

Andolenko, S. (1956). The Imperial Heritage. In B. H. Liddell Hart (Ed.), *The Red Army* (pp. 13–23). New York: Harcourt, Brace and Company.

Aurelius, M. (2002). *The Emperor's Handbook. A New Translation of the Meditations by C. Scot Hicks and David V. Hicks*. New York: Scribner.

Bassford, C. (1994). *Clausewitz in English: The Reception of Clausewitz in Britain and America 1815–1945*. New York/Oxford: Oxford University Press.

Bayerlein, F. (1956). The Armoured Forces. In B. H. Liddell Hart (Ed.), *The Red Army* (pp. 307–312). New York: Harcourt, Brace and Company.

Beaufre, A. (1967). *Strategy of Action*. New York: Praeger.

Bond, B. (1998). *The Pursuit of Victory: From Napoleon to Saddam Hussein*. Oxford: Oxford University Press.

Bucholz, A. (2001). *Moltke and the German Wars, 1864–1871*. New York, NY: Palgrave Macmillan.

Carver, Michael, Field Marshal Lord. (1979). *The Apostles of Mobility. The Theory and Practice of Armoured Warfare*. London: Weidenfield and Nicolson.

Cleary, T. (2005). *The Japanese Art of War - Understanding the Culture of Strategy*. London: Shambhala Classics.

Colby, E. (1943). *Masters of Mobile Warfare*. Princeton: Princeton University Press.

Corum, J. S. (1992). *The Roots of Blitzkrieg - Hans von Seeckt and German Military Reform*. Lawrence: University Press of Kansas.

Danchev, A. (1998). *Alchemist of War – The Life of Basil Liddell Hart*. London: Phoenix Giant.
de Bloch, J. (1914). *Future of War in its Technical Economic and Political Relations*. Boston: The World Peace Foundation.
De Gaulle, C. (1976). *The Army of the Future*. Westport: Greenwood Press.
de Jomini, A. H. (1992). *The Art of War*. London: Greenhill Books.
de Jomini, A. H. (2007). *The Art of War*. Rockville: Arc Manor.
de Saxe, M. (1944). *Reveries upon the Art of War*. Harrisburg: Military Service Pub..
de Saxe, M. (1987). My Reveries upon the Art of War. In T. R. Phillips (Ed.), *Roots of Strategy* (pp. 177–300). Harrisburg: Stackpole Books.
Delbrück, H. (1990). *Warfare in Antiquity - History of the Art of War* (Vol. I). Lincoln/London: University of Nebraska Press.
Delbrück, H. (1997). *Delbrück's Modern Military History*. Edited and Translated by A. Bucholz. Lincoln/London: University of Nebraska Press.
du Picq, A. (1987). Battle Studies – Ancient and Modern Battle. In *Roots of Strategy* (Book 2, pp. 65–300). Harrisburg: Stackpole Books.
Dupuy, T. N. (1987). *Understanding War – History and Theory of Combat*. New York: Paragon House Publishers.
Echevarria, A. J., II. (2013). *Clausewitz and Contemporary War*. Oxford: Oxford University Press.
Ehrfurth, W. (1991). Surprise. In *Roots of Strategy* (Book 3, pp. 351–554). Mechanicsburg: Stackpole Books.
Fergusson, B. (1961). *Wavell – Portrait of a Soldier*. London: Collins.
Finkel, M. (2011). *On Flexibility - Recovery from Technological and Doctrinal Surprise on the Battlefield*. Stanford: Stanford University Press.
Foch, F. (1920). *The Principles of War*. New York: Henry Holt and Company.
Foch, M. F. (1931). *The Memoirs of Marshal Foch Doubleday*. Doran and Company: Garden City.
Franks, T. (2004). *American Soldier*. New York: Regan Books.
Frederick The Great. (1985). Military Instructions for the Generals. In T. R. Phillips (Ed.), *Roots of Strategy* (pp. 301–400). Harrisburg: Stackpole Books.
Fuller, J. F. C. (1923). *The Reformation of War*. London: Hutchinson.
Fuller, J. F. C. (1926). *The Foundations of the Science of War*. London: Hutchinson.
Fuller, J. F. C. (1932). *The Dragon's Teeth – A Study of War and Peace*. London: Constable& Co..
Fuller, J. F. C. (1933). *Generalship: Its Diseases and Their Cure - a Study of the Personal Factor in Command*. London: Faber and Faber.
Fuller, J. F. C. (1943). *Armoured Warfare – An Annotated Edition of Fifteen Lectures on Operations Between Mechanized Forces*. London: Eyre & Spottiswoode.
Fuller, J. F. C. (1948). *The Second World War 1939–45. A Strategical and Tactical History*. London: Eyre & Spottiswoode.

Fuller, J. F. C. (1960). *The Generalship of Alexander The Great*. New Brunswick: Rutgers University Press.
Fuller, J. F. C. (1961). *The Conduct of War 1789–1961: A Study of the Impact of the French, Industrial, and Russian Revolutions on War and Its Conduct*. London: Eyre & Spottiswoode.
Fuller, J. F. C. (1965). *Julius Caesar – Man, Soldier, and Tyrant*. London: Eyre & Spottiswoode.
Gat, A. (2001). *A History of Military Thought from the Enlightenment to the Cold War*. Oxford: Oxford University Press.
Giap, Vo Nguyen. (1962). *People's War People's Army – The Viet Cong Insurrection Manual for Underdeveloped Countries*. New York: Frederick A. Praeger.
Giap, Vo Nguyen. (1970). *Banner of People's War, the Party's Military Line*. London: Pall Mall Press.
Goltz, Field-Marshal Baron von der. (1906). *The Nation in Arms – A Treatise on Modern Military Systems and the Conduct of War*. London: Hodder and Stoughton.
Gray, C. S. (2007). *Fighting Talk – Forty Maxims on War, Peace, and Strategy*. Westport: Praeger Security International.
Guderian, H. (1937). *Armed Forces and Their Cooperation with Other Arms*. Fort Humphreys: The Army War College.
Guderian, H. (1956). *Sotilaan Muistelmia, Org. Erinnerungen Eines Soldaten. Transl. Wolf H. Halsti*. Helsinki: Kustannusosakeyhtiö Otava.
Guderian, H. (1992). *Achtung – Panzer! The Development of Tank Warfare*. (Trans: Christopher Duffy). London: Cassell.
Handel, M. I. (2001). *Masters of War - Classical Strategic Thought*. (3rd Revised and Expanded Edition). London/New York: Routledge.
Hart, R. A. (2006). *Guderian – Panzer Pioneer or Myth Maker*. Washington: Potomac Books.
Howard, M. (1983). *Clausewitz*. Oxford: Oxford University Press.
Isserson, G. S. (2013). *The Evolution of Operational Art*. (Transl.: Bruce W. Menning). Fort Leavenworth: Combat Studies Institute Press.
Kagan, F. W. (2006). *Finding the Target - the Transformation of American Military Policy*. New York: Encounter Books.
Kesseli, P. (2001). *In Pursuit of Mobility: The Birth and Development of Israeli Operational Art: From Theory to Practice*. Sotahistorian Laitos. Publication Series 1 No: 6. Maanpuolustuskorkeakoulu: Helsinki.
Lawrence, T. E. (1935). *Seven Pillars of Wisdom*. London: Jonathan Cape.
Leo VI. (2010). *Taktika*. Text, Translation and Commentary by T. George. Washington: Dennis. Dumbarton Oaks Texts XII.
Leonhard, R. R. (1991). *The Art of Maneuver - Maneuver-Warfare Theory and Air Land Battle*. Novato: Presido.

Leonhard, R. R. (1998). *The Principles of War for the Information Age*. New York: Ballantine Books.

Liddell Hart, B. H. (1927). *The Remaking of Modern Armies*. London: John Murray.

Liddell Hart, B. H. (1932). *When Britain Goes to War: Adaptability and Mobility*. London: Faber and Faber.

Liddell Hart, B. H. (1936). *The War in Outline 1914–1918*. London: Faber and Faber Limited.

Liddell Hart, B. H. (1937). *Europe in Arms*. London: Faber and Faber Limited.

Liddell Hart, B. H. (1948). *The Other Side of the Hill – Germany's Generals, Their Rise and Fall, with Their Own Account of Military Events 1939–1945*. London: Cassell and Co..

Liddell Hart, B. H. (1950). *Defence of the West – Some Riddles of War and Peace*. London: Cassell and Co..

Liu An. (2012). *The Dao of the Military: Liu An's Art of War*. Translated with an Introduction by Andrew Seth Meyer. New York: Columbia University Press.

Longworth, P. (1965). *The Art of Victory – The Life and Achievements of Generalissimo Suvorov 1729–1800*. London: Constable.

Ludendorff, E. (1919). *Sotamuistelmani 1914–1918*. Porvoo: WSOY.

Luttwak, E. N. (1987). *Strategy: The Logic of War and Peace*. Cambridge: Belknap Press of Harvard University Press.

MacArthur, D. (1964). *Reminiscences*. London: Heineman.

Machiavelli, N. (1965). *The Art of War*. A Revised Edition of the Ellis Farneworth Translation. Indianapolis: Bobbs-Merrill Company.

Machiavelli, N. (2003). *The Prince*. Translated by George Bull. London: Penguin Books.

Mao Tse-Tung. (1963). *Selected Military Writings*. Peking: Foreign Languages Press.

Martel, G. (1945). *The Problem of Security*. London: Michael Joseph Ltd..

Matheny, R. R. (2012). *Carrying the War to the Enemy – American Operational Art to 1945*. Norman: Oklahoma University Press.

Montgomery. (1961). *The Path to Leadership*. London: Collins.

Musashi, M. (1995). *Maa, Vesi, Tuli, Tuuli ja Tyhjyys*. (Orig. *Go rin no sho*), Helsinki: Otava.

Patton, G. S., Jr. (1947). *War as I Knew it*. Cambridge: Riverside Press.

Quintus Curtius Rufus. (1988). *The History of Alexander*. New York/London: Penguin Books.

Reid, B. H. (1997). *J. F. C. Fuller: Military Thinker*. London: Macmillan Press.

Reid, B. H. (1998). *Studies in British Military Thought: Debates with Fuller and Liddell Hart*. Lincoln: University of Nebraska Press.

Ritter, G. (1958). *The Schlieffen Plan: Critique of a Myth*. London: Oswald Wolff Publishers.

Rokossovsky, K. (1970). *A Soldier's Duty*. Moscow: Progress Publishers.
Rommel, E. (1953). *The Rommel Papers*. Ed. B.H. Liddell Hart. London: Collins.
Rommel, E. (2003). *Rommel and his Art of War*. Ed. Dr. J. Pimlott. London: Greenhill Books.
Schwarzkopf, N. H. (1993). *It Doesn't Take a Hero – The Autobiography, Written with Peter Petre*. New York, NY: Bantam Books Paperback edition.
Sikorski, General W. (1943). *Modern Warfare – Its Character, its Problems*. London: Hutchinson & Co..
Simpkin, R. E. (1985). *Race to the Swift – Thoughts on Twenty-First Century Warfare*. London: Brassey's Defence Publishers.
Simpkin, R. E. (1987). *Deep Battle - the Brainchild of Marshal Tukhachevskii*. London: Brassey's Defence Publishers.
Smith, R. (2008). *The Utility of Force: The Art of War in the Modern World*. New York: Vintage Books.
Sokolovsky, V. D. (1963). *Military Strategy: Soviet Doctrine and Concepts*. London/Dunmow: Pall Mall Press.
Stein, M. (2007). *Field Marshal Von Manstein, a Portrait – The Janus Head*. Solihull: Helion.
Stempel, J. (2012). *The Nature of War – Origins and Evolution of Violent Conflict*. Jefferson: McFarland & Co.
Strachan, H. (2007). Clausewitz and the Dialectics of War. In H. Strachan & A. Herberg-Rothe (Eds.), *Clausewitz in the Twenty-First Century* (pp. 14–44). Oxford: Oxford University Press.
Strachan, H. (2011). Operational Art and Britain, 1909-2009. In J. A. Olsen & M. van Creveld (Eds.), *The Evolution of Operational Art – From Napoleon to the Present* (pp. 96–136). Oxford: Oxford University Press.
Strachan, H. (2013). *The Direction of War – Contemporary Strategy in Historical Perspective*. Cambridge: Cambridge University Press.
Sun Tzu. (1971). *The Art of War*. Translated and with an Introduction by B. Samuel. Griffith with a Foreword by B.H. Liddell Hart. New York/Oxford: Oxford University Press.
Svechin, A. A. (1992). *Strategy*. Boulder: Eastview Press.
Thucydides. (1971). *The Peloponnesian War*. (Transl.: Rex Warner). Harmondsworth, Middlesex: Penguin Books.
Toffler, A. (1990a). *The Third Wave*. New York: Bantam Books, Paperback edition..
Toffler, A. (1990b). *Future Shock*. New York: Bantam Books, Paperback edition.
Toffler, A., & Toffler, H. (1995). *War and Anti-War: Survival at the Dawn of the 21st Century*. New York: Warner Books.
Tsunetomo, Y. (n.d.). *Hagakure, Book of the Samura*. Ebook, Downloaded iTunes Store 15.5.2011.
van Creveld, M. (2008). *The Culture of War*. New York: Ballantine Books.

Vego, M. (2009). *Joint Operational Warfare: Theory and Practice*. (Revised Ed.). Newport: U.S. Naval War College.
von Bernhardi, F. (1914). *How Germany Makes War*. London: Hodder & Stoughton.
von Bülow, D. H. (2013). *The Spirit of the Modern System of War*. Cambridge: Cambridge University Press.
von Clausewitz, C. (1989). *On War*. Edited and Translated by Michael Howard and Peter Paret. First paperback edition. Princeton: Princeton University Press.
von Freytag-Loringhoven, H. (1991). The Power of Personality in War. In *Roots of Strategy* (Book 3, pp. 173–349). Mechanicsburg, Pennsylvania: Stackpole Books.
von Manstein, E. (1982). *Lost Victories*. Novato: Presido Press.
von Moltke, H. (1993). *On the Art of War – Selected Writings*. Ed. Daniel J. Hughes. New York: Ballantine Books.
von Schlieffen, Count Alfred. (1936). *Cannae*. Fort Leavenworth: The Command and General Staff School Press.
von Seeckt, H. (1930). *Valtiomies ja Sotapäällikkö*. Helsinki: Otava.
Wavell, A. (1941). *Generals and Generalship*. New York: Macmillan.
Wavell, A. (1953). *Soldiers and Soldiering or Epithets of War*. London: Jonathan Cape.
Yuen, D. M. C. (2014). *Deciphering Sun Tzu - How to Read the Art of War*. Oxford: Oxford University Press.

Index

J. Hanska, *War of Time*,
https://doi.org/10.1007/978-3-030-45517-0

O

P

The manufacturer's authorised representative in the EU is Springer Nature Customer Service Centre GmbH, Europaplatz 3, 69115 Heidelberg, Germany. If you have any concerns regarding our products, please contact ProductSafety@springernature.com

Printed and bound by CPI Group (UK) Ltd, Croydon, CR0 4YY

30/09/2026

02225644-0013